# Java语言程序设计案例教程

主　编　杨韵芳　杨克松

参　编　王燕红　骆惠清　王国栋　郑小娟

骆旭坤　陈纯纯　龚清湄　杨　岚

北京理工大学出版社

BEIJING INSTITUTE OF TECHNOLOGY PRESS

## 内 容 简 介

本书从初学者角度出发，通过丰富的案例讲述 Java 语言的理论知识及编程方法，内容涵盖 Java 编程基础，面向对象封装性、继承性及多态性，图形用户界面，集合类，字符串类，Lambda 表达式与 Stream，异常处理，IO 流，多线程编程，JDBC 编程，网络编程等。读者可以跟随本书的讲解，结合案例，理实结合，设计中小型应用程序。

本书可作为计算机、软件工程、物联网、网络工程、人工智能、信息计算等专业的教学用书，也可作为相关领域的培训教材、参考用书。



**图书在版编目（CIP）数据**

Java 语言程序设计案例教程 / 杨韵芳，杨克松主编.
北京：北京理工大学出版社，2025. 1.
ISBN 978 -7 -5763 -4688 -6

Ⅰ. TP312. 8

中国国家版本馆 CIP 数据核字第 2025SJ3354 号

**责任编辑**：钟　博　　**文案编辑**：钟　博
**责任校对**：刘亚男　　**责任印制**：施胜娟

**出版发行** / 北京理工大学出版社有限责任公司
**社　　址** / 北京市丰台区四合庄路 6 号
**邮　　编** / 100070
**电　　话** / (010) 68914026（教材售后服务热线）
(010) 63726648（课件资源服务热线）
**网　　址** / http://www.bitpress.com.cn

**版 印 次** / 2025 年 1 月第 1 版第 1 次印刷
**印　　刷** / 河北盛世彩捷印刷有限公司
**开　　本** / 787 mm × 1092 mm　1/16
**印　　张** / 22. 25
**字　　数** / 493 千字
**定　　价** / 89. 00 元

**图书出现印装质量问题，请拨打售后服务热线，负责调换**

# 前言

随着信息技术的飞速发展，Java 编程语言在全球范围内展现出强大的生命力，广泛应用于互联网、移动开发、大数据等诸多前沿领域。为了帮助不同层次的读者——无论是高等院校、职业院校的在校生，还是对编程怀有热情的零基础初学者——能顺利掌握 Java 编程技能，我们编写了本书。

本书以实用为出发点，构建了一套循序渐进的学习体系。开篇通过趣味盎然的入门案例，引领读者认识 Java 世界，熟悉 Java 开发环境的搭建，为后续的学习筑牢根基。全书摒弃了传统枯燥的理论堆砌模式，代之以丰富、实用且贯穿始终的案例讲解模式——从基础语法的入门示例，到复杂项目的实战演练，每个案例都经过精心设计，旨在让读者在实践中领悟 Java 编程的精髓，轻松跨越从理论知识到实际应用的鸿沟。

本书采用以项目为驱动、问题分解的前沿教学方法，以达到简化复杂问题的目的，让初学者更易学习和掌握相关知识，全书的任务完成后形成若干小项目，经过编者多年的实践经验，这种系统化的学习效果是比较理想的。

本书在内容选取及组织上，以学生为中心、以岗位职业能力为目标、以实用性为原则、以项目为载体，通过丰富的案例讲述 Java 理论知识及编程方法，内容涵盖 Java 编程基础，面向对象封装性、继承性及多态性，图形用户界面（GUI），集合类，字符串类、Lambda 表达式与 Stream，异常处理，IO 流，多线程编程，JDBC 编程及网络编程等知识点。

本书第 2、3、5、6 章由杨韵芳编写，全书任务工单由郑小娟编写，第 1、4、7 章由杨克松编写，第 8、9 章由王燕红、王国栋编写，第 10 章由骆惠清编写，第 11 章由陈纯纯、骆旭坤编写，第 12 章由龚清湄、杨岚编写。本书的统稿、定稿工作由杨克松、杨韵芳、王燕红完成。

为了对接新一代信息技术的发展趋势和产业需求，我们与福建国科信息科技有限公司深度合作，企业工程师全程参与课程调研、岗位需求分析、教材内容及项目确定、任务划分、任务工单设计，以及教材编写的全过程，准确对接软件开发岗位职业能力需求。

本书的编写还得到了黎明职业大学各级领导及广大教师的支持和协助，在此表示由衷的感谢。

衷心希望读者借助本书，顺利踏入 Java 编程的精彩世界。

由于编者水平有限，书中难免存在疏漏，恳请各位专家和读者批评指正。

**编　者**

# 目录

第 1 章

# Java 开发入门

——千里之行，始于足下

- 1.1 Java开发环境搭建
- 1.2 Java基础语法

# 1.1 Java 开发环境搭建

## 学习目标

（1）掌握 Java 开发环境的搭建方法。

（2）掌握 Eclipse 的使用方法。

（3）掌握 Java 程序的基本结构。

## 相关知识

### 1.1.1 JRE 与 JDK

JRE（ Java Runtime Environment）是 SUN 公司提供的 Java 运行环境，只包含 Java 运行工具，不包含 Java 编译器，专为普通用户准备。

JDK（ Java Development Kit）是 SUN 公司提供的 Java 开发环境，是 Java 的核心开发包，包括 Java 编译器、Java 调试器、Java 运行工具、Java 打包工具等。JDK 自带 JRE，因此，开发人员只需要再安装 JDK 即可。

### 1.1.2 JDK 的安装及配置

#### 一、下载 JDK

Oracle 公司提供了支持多种操作系统的 JDK，用户可以根据自己的操作系统选择指定版本的 JDK 安装包，从官网 https：// www. oracle. com/java/technologies/downloads/进入图 1－1－1 所示界面，下载 Window 64 位操作系统 JDK21 版本安装包“jdk－21_ windows－x64_ bin. exe”。

Java downloads　Tools and resources　Java archive

JDK 21 binaries are free to use in production and free to redistribute, at no cost, under the Oracle No-Fee Terms and Conditions (NFTC).

JDK 21 will receive updates under the NFTC, until September 2026, a year after the release of the next LTS. Subsequent JDK 21 updates will be licensed under the Java SE OTN License (OTN) and production use beyond the limited free grants of the OTN license will require a fee.

Linux　macOS　Windows

| Product/file description | File size | Download |
| --- | --- | --- |
| x64 Compressed Archive | 185.52 MB | https://download.oracle.com/java/21/latest/jdk-21_windows-x64_bin.zip (sha256) |
| x64 Installer | 163.91 MB | https://download.oracle.com/java/21/latest/jdk-21_windows-x64_bin.exe (sha256) |
| x64 MSI Installer | 162.07MB | https://download.oracle.com/java/21/latest/jdk-21_windows-x64_bin.msi (sha256) |

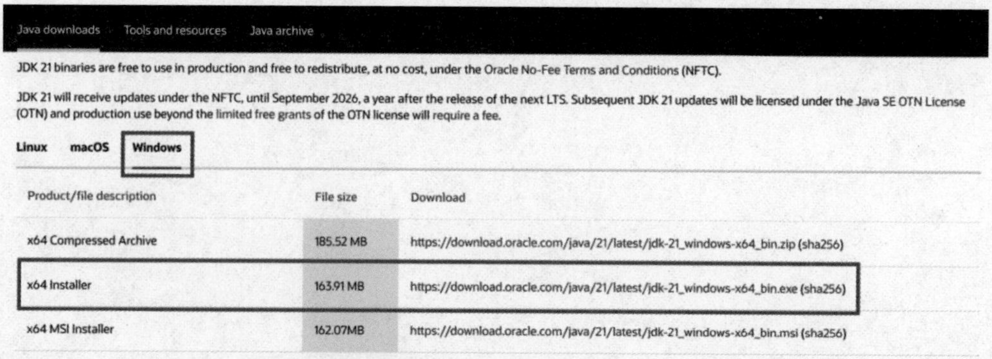

图 1－1－1　Window 64 位操作系统 JDK21 版本安装包下载

（1）x64 Compressed Archive：x64 架构压缩包，免安装，解压后可以直接使用，但需要手动安装 JRE。

（2）x64 installer：x64 架构 EXE 安装包，自动配置。

（3）x64 msi installer：x64 架构 MSI 安装包，自动配置。

## 二、安装 JDK

（1）运行“jdk - 21_windows - x64_bin. exe”。

（2）设置安装路径。默认安装路径为“C:\Program Files\Java\jdk - 21”，也可以通过“更改”按钮修改安装路径，如图 1 - 1 - 2 所示。

（3）安装结束，单击“关闭”按钮。

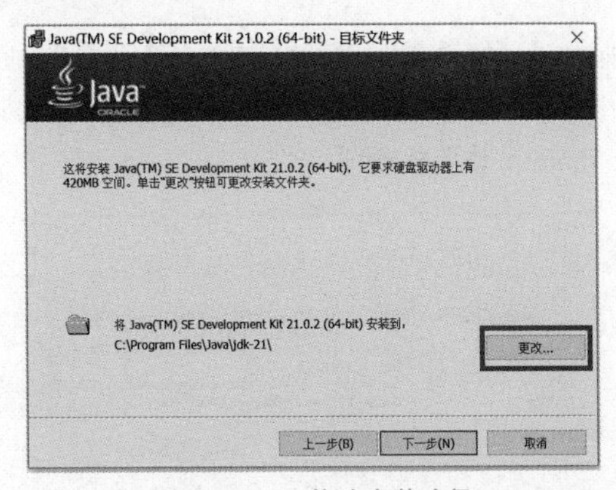


图 1 - 1 - 2　修改安装路径

## 三、配置环境变量

（1）在“开始”菜单中选择“环境变量”→“编辑系统环境变量”命令。

（2）打开“系统属性”对话框，在“高级”选项卡中单击“环境变量”按钮，如图1 - 1 - 3 所示。

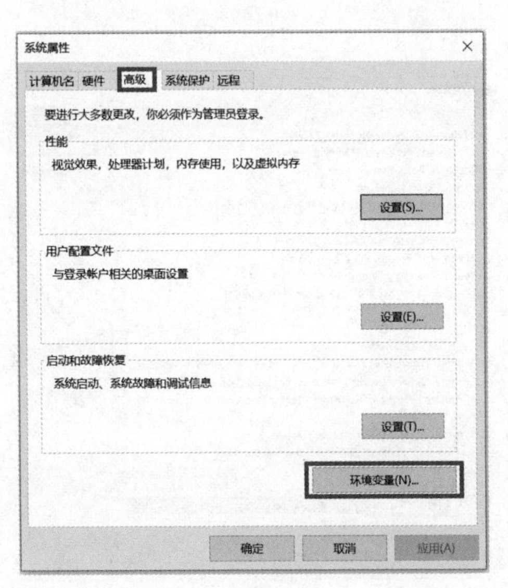


图 1 - 1 - 3　“系统属性”对话框

（3）新建系统变量 JAVA_HOME。选择“系统变量”→“新建”命令，打开图 1－1－4 所示对话框，输入变量名“JAVA_HOME”，变量值为 JDK 安装路径。

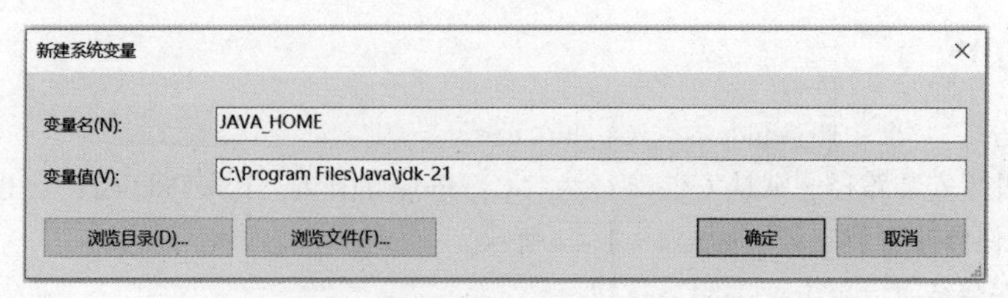


图 1－1－4　新建系统变量 JAVA_HOME

（4）在图 1－1－5 所示“环境变量”对话框中“系统变量”列表框中选择 Path 变量后单击“编辑”按钮，打开图 1－1－6 所示“编辑环境变量”对话框，单击“新建”按钮，输入“%JAVA_HOME%\bin”，并上移置顶。

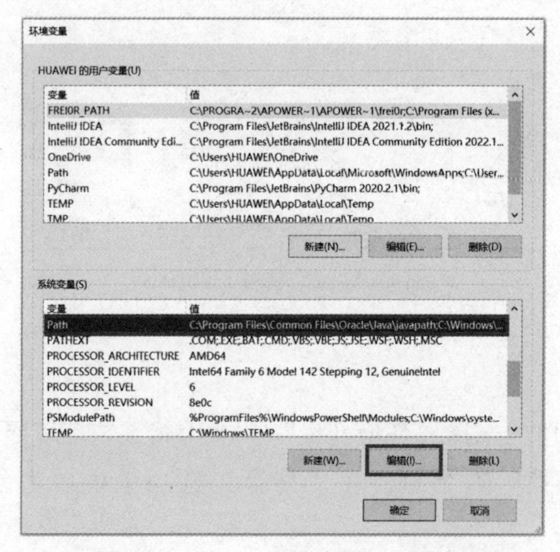


图 1－1－5　“环境变量”对话框

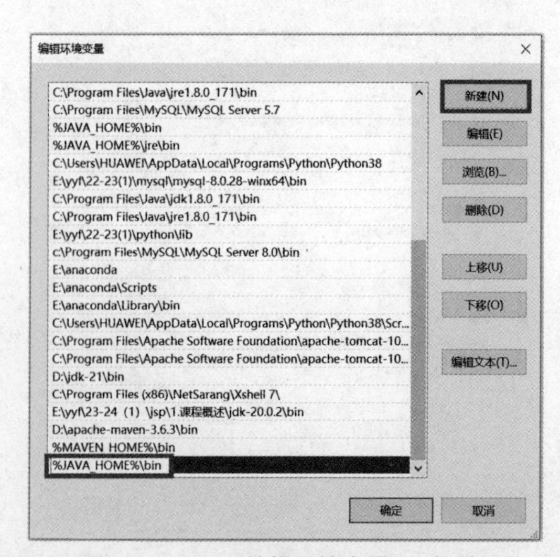


图 1－1－6　“编辑环境变量”对话框

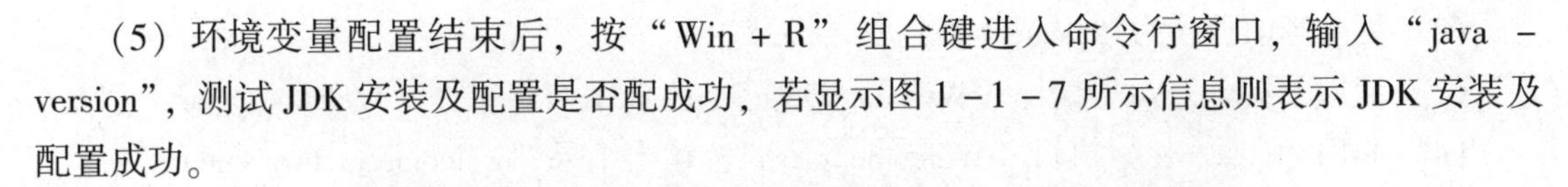

（5）环境变量配置结束后，按“Win + R”组合键进入命令行窗口，输入“java -version”，测试 JDK 安装及配置是否配成功，若显示图 1－1－7 所示信息则表示 JDK 安装及配置成功。

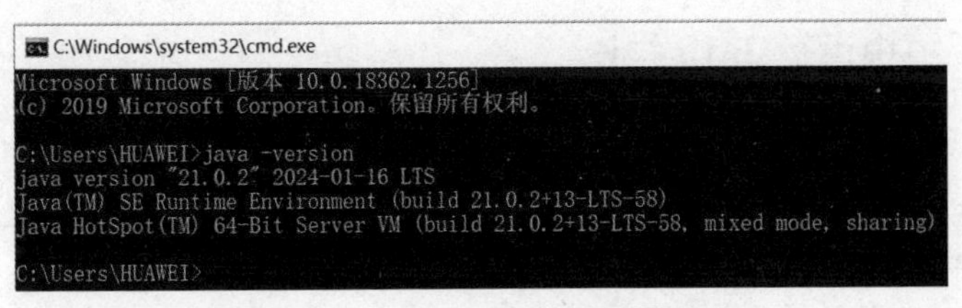


图 1－1－7　JDK 安装及配置成功

## 1.1.3　开发工具

### 一、在 DOS 环境中运行 Java 程序

**1. 编写 Java 源文件**

在 JDK 安装目录的 bin 目录下新建文本文件并重命名为“Welcome. java”，使用记事本打开“Welcome. java”，编写第一个 Java 程序，代码如下：

```
1   class Welcome {
2       public static void main(String[ ] args){
3       System. out .println("Welcome to Java code!");
4       }
5   }
```

【说明】

程序第 1 行使用 class 关键字定义一个类，类名为 Welcome。在 Java 中，类相当于一个程序，所有代码都写在类定义的大括号中。第 2 行在类中定义了程序的入口 main( ) 方法，主函数需要执行的代码写在 main( ) 方法所属大括号中。第 3 行编写了一条输出语句“System. out. println ( " welcome to Java code !");”。运行程序时将在命令行窗口输出“welcome to Java code ! ”。

**2. 运行“Welcome. java”**

（1）进入命令行窗口，如图 1－1－8 所示。

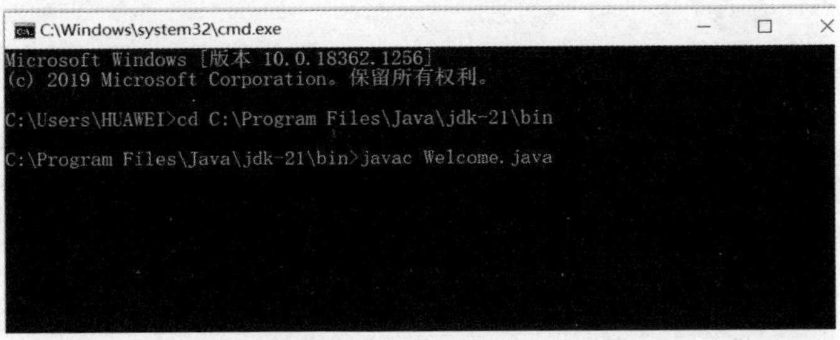


图 1－1－8　命令行窗口

（2）更改当前目录为“javac. exe”所在目录。

（3）调用“javac. exe”编译“Welcome. java”，生成字节码文件“Welcome. class”。

（4）调用“java. exe”执行“Welcome. class”，输出结果“welcome to Java code !”。

## 二、在 Eclipse 环境中运行 Java 程序

### 1. 常用 Java 开发工具

（1）Eclipse（推荐）。下载地址：http：//www. eclipse. org/downloads/packages/。

（2）IDEA。下载地址：https：//www. jetbrains. com/idea/download/。

（3）VSCode（全称：Visual Studio Code）。下载地址：https：//code. visualstudio. com/download。

### 2. 下载 Eclipse

（1）进入官方网站（http：//www. eclipse. org），单击“Download”按钮进行免费下载，如图 1－1－9 所示。本书使用的 Eclipse 版本是“eclipse－jee－2023－06”。

图 1－1－9　Eclipse 官网

（2）选择 Eclipse 下载入口“Download Packages”，如图 1－1－10 所示。

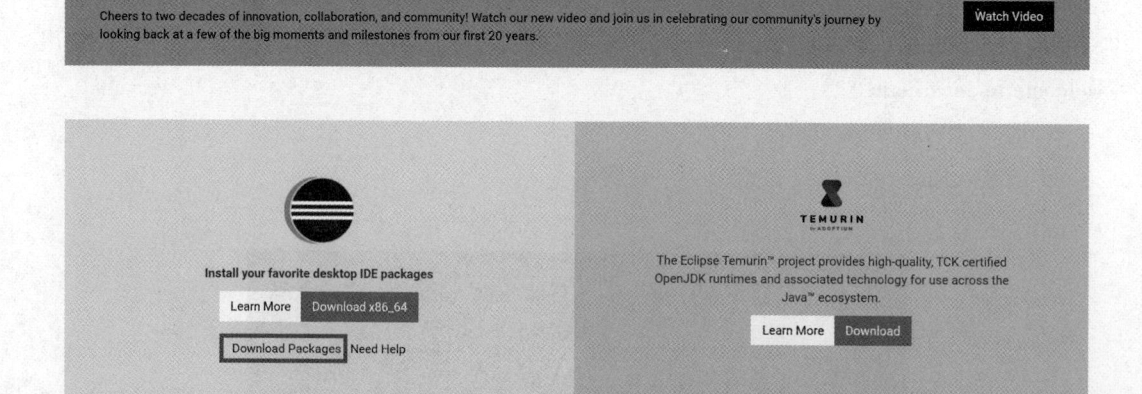


图 1－1－10　选择 Eclipse 下载入口

（3）如图 1－1－11 所示，选择“2023－12”版本。

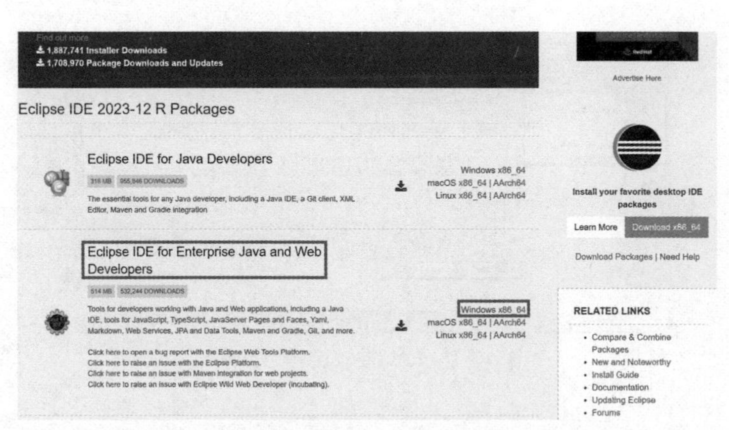


图 1－1－11　选择“2023－12”版本

（4）单击“Download”按钮，若下载速度低，可选择“Select Another Mirror”命令，选择一个国内的镜像地址下载。

### 3. 使用 Eclipse

Eclipse 是一个免安装的软件，解压后即可使用。

（1）设置工作空间（Workspace）。

工作空间用于保存 Eclipse 中创建的项目和相关设置。可以使用 Eclipse 提供的默认路径作为工作空间，也可以单击“Browse”按钮来更改路径，工作空间设置完成后，单击“OK”按钮即可。

（2）设置工作空间所使用字符集为“UTF－8”。

在“Window”菜单中选择“Preferences”选项，打开图 1－1－12 所示“Preferences”对话框。

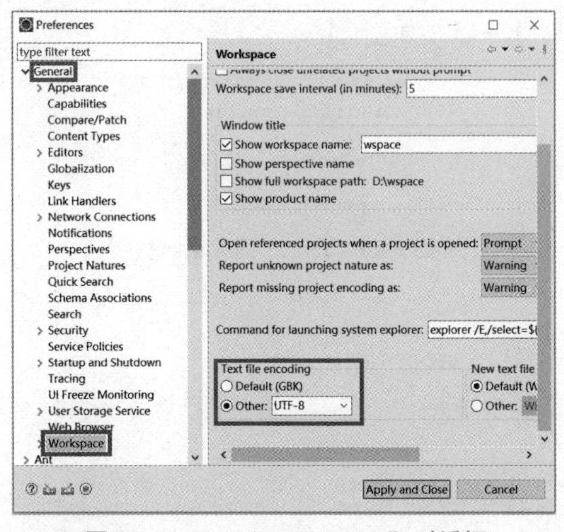


图 1－1－12　“Preferences”对话框

（3）新建工程。

在“File”菜单中选择“New”→“Project”→“Java Project”选项。

（4）新建“Welcome. java”并运行，控制台输出结果如图 1－1－13 所示。

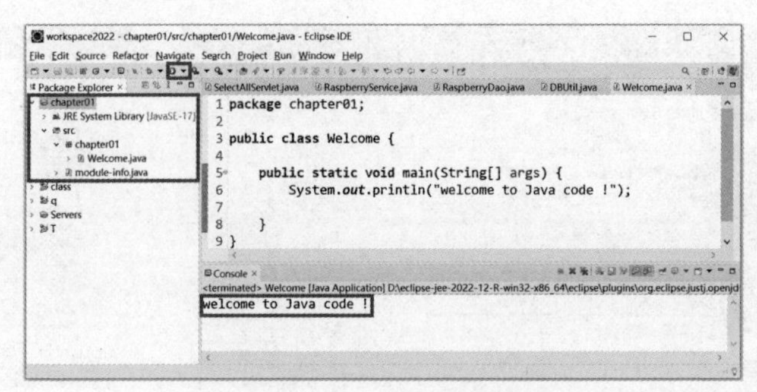

图 1-1-13　控制台输出结果

### 1.1.4　Java 的工作原理

如图 1-1-14 所示，Java 程序运行流程经过以下 3 个步骤。

（1）编写 Java 源文件：在 Java 开发环境中编写代码，保存为后缀名为 .java 的 Java 源文件。

（2）编译：Java 编译器对 Java 源文件进行编译，生成后缀名为 .class 的字节码文件。

（3）运行：Java 解释器将字节码文件翻译成机器代码并运行，显示结果。

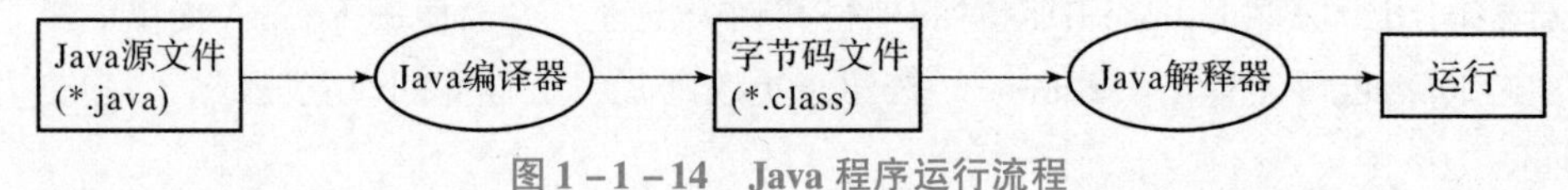


图 1-1-14　Java 程序运行流程

**案例**

创建"class01"工程，定义 Calculate 类，主函数实现计算两数之和并输出的功能。

**1. 工程目录结构**

工程目录结构如图 1-1-15 所示。

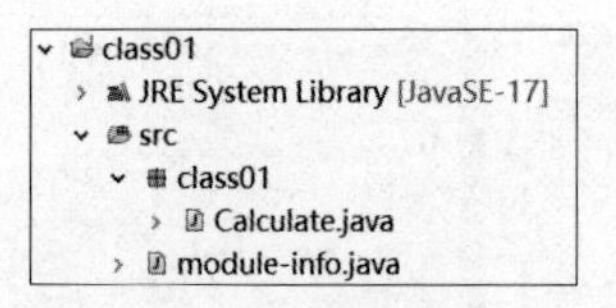


图 1-1-15　工程目录结构

**2. 程序**

代码如下：

```
1  package class01;
2  public class Calculate {
3      public static void main(String[] args) {
4          int a = 2,b = 3;
5          int s = a + b;
6          System. out .println( "s = " + s);
7      }
8  }
```

## 任务工单

<table>
<tr><td>任务名称</td><td colspan="5">Java 环境搭建</td></tr>
<tr><td>班级</td><td></td><td>学号</td><td></td><td>姓名</td><td></td></tr>
<tr><td>任务要求</td><td colspan="5">（1）下载并安装 JDK21。<br>（2）下载并解压 Eclipse 2023。<br>（3）新建“ex01”工程，创建 Test 类，主函数输出“This is my Exercise”。</td></tr>
<tr><td>任务实现</td><td colspan="5"></td></tr>
<tr><td>异常及待解决问题记录</td><td colspan="5"></td></tr>
</table>

# 1.2 Java 基础语法

**学习目标**

（1）了解 Java 程序的基本结构。

（2）掌握 Java 的标识符与关键字。

（3）掌握 Java 程序的调试及纠错方法。

**相关知识**

## 1.2.1 Java 程序的基本规范

### 一、Java 程序的基本结构

Java 程序通过类来组织，也就是说代码必须写在类中。下面以“Welcome. java”为例，分析其基本结构。代码如下：

```
1  public class Welcome {
2      public static void main (String[ ] args) {
3          System. out .println( "Welcome to Java!");
4      }
5      public void func1(){
6          int a =2,b =3;
7          int s = a + b;
8      }
9  }
10  class SayHi{
11      public void hi(){
12          System. out .println( " Hi");
13      }
14  }
```

运行结果如图 1 –2 –1 所示。

```
Welcome to Java!
```

图 1 –2 –1　运行结果

【说明】

（1）一个“. java”文件允许定义多个类（Welcome 类、SayHi 类），但是以 public 修饰的类只能有一个（Welcome 类），并且必须与“. java”文件（“Welcome. java”）同名。

（2）一个类可以定义多个方法，例如 Welcome 类包含 main( )方法、func1( )方法，main( )方法为整个工程的入口。

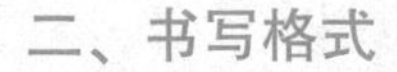

## 二、书写格式

（1）Java 严格区分大小写。例如，定义类时，不能将 class 写成 Class，否则编译会报错。

（2）字符必须在英文半角状态下输入，否则编译器报“illegal character”（非法字符）的错误信息。

（3）Java 程序中一个连续的字符串不能分开在两行中书写。例如：

```
System. out .println ("welcome to Java code!");
```

如果为了便于阅读，将一个字符串分在两行中书写，则可以先将这个字符串分成两个字符串，然后用“+”将这两个字符串连接起来，例如上面的语句可以改写如下：

```
System.out .println ("welcome to" +
                     "Java code!");
```

（4）注释语句是对程序的某个功能或者某行代码的解释说明，它只在 Java 源文件中有效，在编译程序时，编译器会忽略这些注释语句，不会将其编译到字节码文件中。Java 中有 3 种类型的注释语句，具体如下。

①单行注释。

单行注释用于对程序中的某行代码进行注释，用符号“//”表示，“//”后面为被注释的内容。例如：

```
int c = 10;//定义一个整型变量
```

②多行注释。

注释的内容可以为多行，它以符号“/*”开头，以符号“*/”结尾。例如：

```
/* int c = 10;
int x = 5; */
```

③文档注释。

文档注释以符号“/**”开头，直到符号“**/”结束，其中的内容均作为注释，可占多行，一般放在一个变量或一个函数的说明之前。例如：

```
/**
*1. 输入两个数
*2. 求出最大值并输出
**/
```

（5）分隔符。Java 常用分隔符如表 1-2-1 所示。

表 1-2-1　Java 常用分隔符

| 分隔符 | 名称 | 说明 |
| --- | --- | --- |
| {} | 大括号（花括号） | 用于定义块、类、方法及局部范围，也用于括起初始化的数组的值。大括号必须成对出现 |
| [] | 中括号（方括号） | 用于进行数组的声明，也用于撤销对数组值的引用 |
| () | 小括号（圆括号） | 在定义和调用方法时，用于容纳参数表。在控制语句或强制类型转换的表达式中用于表示执行或计算的优先权 |
| ; | 分号 | 用于表示一条语句的结束。语句必须以分号结束，否则即使一条语句跨行或者有多行，也仍是未结束的 |
| , | 逗号 | 在变量声明中用于分隔变量表中的各变量。在 for 控制语句中，用于将小括号里的语句连接起来 |
| : | 冒号 | 用于说明语句标号 |
| . | 圆点 | 用于类、对象及其属性或者方法之间的分隔 |

在 Java 中，除了表 1-2-1 所示的 Java 常用分隔符外，还包括空格、回车、换行和制表符（Tab 键）等空白分隔符，可以通过它们进行适当的排版，使程序整齐美观、层次清晰，提高程序的可读性。Eclipse 提供了一种简单的快速调整程序格式的功能，可以选择“Source”→“Format”命令调整程序格式。如果程序没有错误，则程序格式会变成预定义的样式。在程序编写完成后，进行快速格式化，可以使程序美观整齐，应该养成这个习惯。

## 1.2.2　Java 的标识符

### 一、关键字

关键字是编程语言中事先定义好并被赋予了特殊含义的单词。与其他编程语言一样，Java 中预留了许多关键字，如表 1-2-2 所示。

表 1-2-2　Java 关键字

| abstract | continue | for | new | switch |
| --- | --- | --- | --- | --- |
| assert | default | goto | package | synchronized |
| boolean | do | if | private | this |
| break | double | implements | protected | throw |
| byte | else | import | public | throws |
| case | enum | instanceof | return | transient |

续表

| catch | extends | int | short | try |
|---|---|---|---|---|
| char | final | interface | static | void |
| class | finally | long | strictfp | volatile |
| const | float | native | super | while |

## 二、标识符命名规范

在编程中，经常需要在程序中定义一些名称，例如包名、类名、方法名、参数名、变量名等，这些名称被称为标识符。在 Java 中定义标识符需要注意如下几点。

（1）所有标识符都应该以字母（A ~ Z 或者 a ~ z）、美元符（$）或者下划线(_)开始。

（2）首字符之后可以是字母（A ~ Z 或者 a ~ z）、美元符（$）、下划线(_)或者数字的任何字符组合。

（3）关键字不能用作标识符。

（4）标识符区分大小写。

①合法标识符举例：age、$ salary、_value、__1_value。

②非法标识符举例：123abc、－salary、case。

## 三、标识符业内规则

定义标识符时必须严格遵守上述标识符命名规范，否则程序在编译时会报错。除了上面列出的标识符命名规范外，为了增强程序的可读性，定义标识符时业内默认遵循以下规则。

（1）包名的所有字母一律小写，例如 cn. lmu. chaptor 01。

（2）类名和接口名的每个单词的首字母都大写，例如 Array、List、lterator。

（3）常量名的所有字母都大写，单词之间用下划线连接，例如 DAY_OF_MONTH。

（4）变量名和方法名的第一个单词的首字母小写，从第 2 个单词开始，每个单词的首字母大写，例如 maxSalary、selectUserById。

（5）在程序中，应该尽量使用有意义的单词来定义标识符，以使程序便于阅读，例如使用 userName 表示用户名，使用 password 表示密码。

## 任务工单

<table>
<tr><td>任务名称</td><td colspan="5">错误检测及代码调试</td></tr>
<tr><td>班级</td><td></td><td>学号</td><td></td><td>姓名</td><td></td></tr>
<tr><td>任务要求</td><td colspan="5"><pre>public class detect {
    public static void main(string[ ] args) {
        int a =2,b =3;
        int s = a + b;
        System. out .println("s = " + s)_
    }
}</pre>（1）观察并发现上述程序中的 3 个错误。<br>（2）运行程序，根据错误提示，修改 3 个错误。</td></tr>
<tr><td>任务实现</td><td colspan="5"></td></tr>
<tr><td>异常及待<br>解决问题记录</td><td colspan="5"></td></tr>
</table>

## 练习题

### 1. 选择题

（1）main( )方法是 Java 程序执行的入口。关于 main( )方法，以下合法的是（　　）。

A. Public static void main( )

B. Public static void main （String args[ ]）

C. public static int main （String [ ]arg）

D. public void main （String arg[ ]）

（2）编译 Java 源文件产生的字节码文件的扩展名为（　　）。

A. . Java　　B. . class　　C. . html　　D. . exe

（3）下列说法中正确的是（　　）。

A. Java 是不区分大小写的

B. Java 以方法为程序的基本单位

C. 类名用关键字 class 定义

D. Java 源文件名和类名不允许相同

（4）（　　）是 Java 中不合法的标识符。

A. Spersons　　B. twoNum　　C. _myVar　　D. * point

（5）下列 Java 注释正确的是（　　）。

A. /*我爱北京天安门*/　　B. //我爱北京天安门*/

C. /**我爱北京天安门*/　　D. /*我爱北京天安门**/

（6）下面属于 Java 关键字的是（　　）。

A. For　　B. for　　C. FOR　　D. true

（7）下面属于 Java 语言关键字的是（　　）。

A. null　　B. false　　C. length　　D. instanceof

### 2. 编程题

在 Eclipse 中创建项目并编写程序，实现题图 1－1 所示的输出效果。

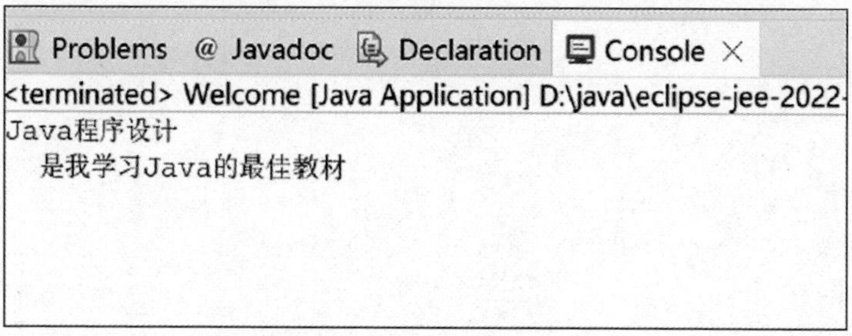

Problems　@ Javadoc　Declaration　Console ×

<terminated> Welcome [Java Application] D:\java\eclipse-jee-2022-

Java程序设计

　是我学习Java的最佳教材

题图 1－1　编程题图

# 第2章

# Java编程基础

——精诚所至，金石为开

# 2.1 Java 数据类型

## 学习目标

（1）了解 Java 数据类型。

（2）掌握变量数据类型的合理定义。

## 相关知识

### 2.1.1 数据类型

数据在计算机中总是以某种特定的格式存放在计算机的存储器中。Java 数据类型分为两大类：8 个基本类型（primitive type）和 5 个引用类型（reference type），完整分类如图 2-1-1 所示。

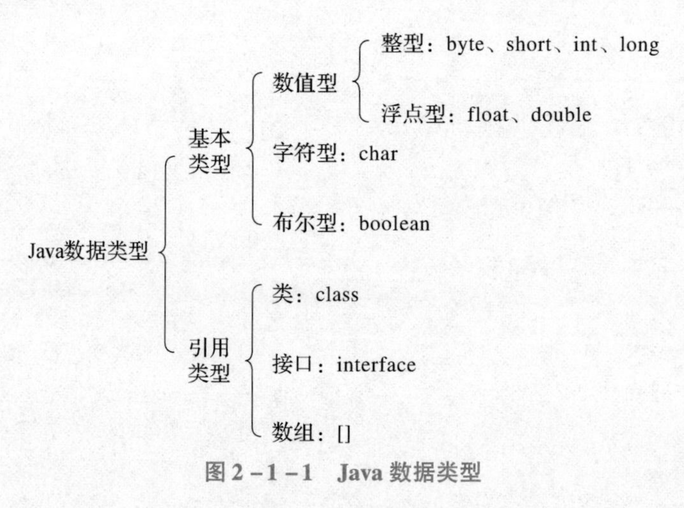


图 2-1-1 Java 数据类型

## 一、基本类型

基本类型是 Java 中预定义的数据类型，由相应的保留关键字表示，具有明确的取值范围和数学行为。基本类型的数据都是单值，不是复杂对象，因此基本类型并不是面向对象的，这主要是出于效率方面的考虑。但是，Java 为基本类型提供了对应的对象版本，即基本类型的包装类（wrapper），具体如表 2-1-1 所示。

## 二、引用类型

引用类型包括类、接口、数组、枚举，后续章节将对此展开介绍。

表 2-1-1　基本类型及其包装类对应关系

| 基本类型 | 占用字节 | 默认值 | 包装类 |
|---|---|---|---|
| char（字符型） | 1 | \ u0000 | Character |
| byte（字节型） | 1 | 0 | Byte |
| short（短整型） | 2 | 0 | Short |
| int（整型） | 4 | 0 | Integer |
| long（长整型） | 8 | OL | Long |
| float（单精度浮点型） | 4 | 0 | Float |
| double（双精度浮点型） | 8 | 0. 0d | Double |
| boolean（布尔型） | 1 | False | Boolean |

## 2.1.2　Java 中的常量

常量是以字面形式直接给出的值，或者以关键字 final 定义的标识符常量。常量一经建立，在程序的整个运行过程中其值不会改变。Java 中的常量按其数据类型分为整型常量、浮点型常量、字符型常量、字符串型常量、布尔型常量和 null 常量。

### 一、整型常量

整型常量是整数类型的数据，有二进制、八进制、十进制和十六进制 4 种表示形式。

### 二、浮点型常量

浮点型常量就是数学中的小数，分为 float（单精度浮点型）和 double（双精度浮点型）两种。其中，单精度浮点型常量以 F 或 f 结尾，而双精度浮点型常量则以 D 或 d 结尾。当浮点型常量不加任何后缀时，Java 虚拟机会默认为双精度浮点型常量。浮点型常量通常有以下两种表示形式。

（1）小数点形式。它由数字和小数点组成，例如 3. 9，-0. 23，-. 23，. 23，0. 23。

（2）指数形式。例如，2. 3e3，2. 3E3 都表示 $2.3\times10^{3}$；. 2e-4 表示 $0.2\times10^{-4}$。

### 三、字符型常量

字符型常量用于表示一个字符。字符型常量以一对英文半角格式的单引号(' ')作为定界符，它可以是英文字母、数字、标点符号以及由转义序列表示的特殊字符。Java 中的转义字符如表 2-1-2 所示。

表 2-1-2　Java 中的转义字符

| 转义字符 | 意义 | ASCII 码值（十进制） |
| --- | --- | --- |
| \ a | 响铃（BEL） | 007 |
| \ b | 退格（BS），将当前位置移到前一列 | 008 |
| \ f | 换页（FF），将当前位置移到下页开头 | 012 |
| \ n | 换行（LF），将当前位置移到下一行开头 | 010 |
| \ r | 回车（CR），将当前位置移到本行开头 | 013 |
| \ t | 水平制表（HT），跳到下一个 Tab 位置 | 009 |
| \ v | 垂直制表（VT） | 011 |
| \ | 代表一个反斜线字符 | 092 |
| ’ | 代表一个单引号（撇号）字符 | 039 |
| “ | 代表一个双引号字符 | 034 |
| \ 0 | 空字符（null） | 000 |
| \ ddd | 1～3 位八进制数所代表的任意字符 | 3 位八进制 |
| \ xhh | 1～2 位十六进制所代表的任意字符 | 3 位八进制 |

## 四、字符串型常量

字符串型常量用于表示一串连续的字符。字符串型常量以一对英文半角格式的双引号（""）作为定界符。字符串型常量以对象的形式出现，在 java. lang 包中 String 类的实例（对象）就是字符串型常量，字符串型常量可以包括转义字符，例如" Hello"、" two \nlines"、" \22\ u9f7\n K 1234\n"。

## 五、布尔型常量

布尔型常量即布尔型的两个值 true 和 false，其用于区分一个事物的真与假。

## 六、null 常量

null 常量只有一个值 null，表示对象的引用为空。null 可以赋给任意引用类型或者转化成任意引用类型。注意，null 常量不等同于数值 0，也不等同于空字符串。

### 2.1.3　Java 中的变量

## 一、变量的定义

变量是 Java 程序中的基本存储单元。在 Java 中变量必须先声明后使用，声明变量包括给出变量的名称和指明变量的数据类型，必要时还可以指定变量的初始值，其语法格式如下：

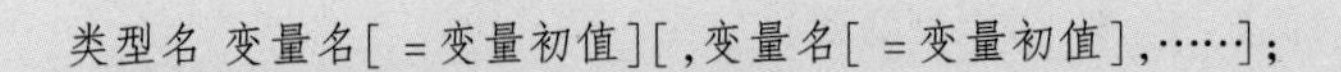

```
类型名 变量名[ =变量初值][,变量名[ =变量初值],……];
```

例如：

```
int x=0,y;
y=x+3;
```

在上面的代码中，第 1 行代码定义了两个变量 x 和 y，计算机为它们分配了两个内存单元，并初始化变量 x 为 0，而没有为变量 y 分配初始值。执行第 2 行代码时，程序首先取出变量 x 的值，与 3 相加后，将结果赋给变量 y，此时变量 x 和 y 在内存中的状态发生了变化，如图 2－1－2 所示。

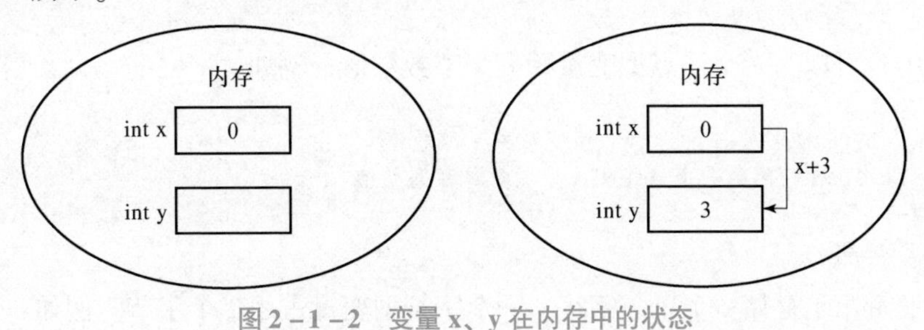


图 2－1－2　变量 x、y 在内存中的状态

## 二、变量的数据类型

### 1. 整型变量

整型变量用来存储整数数值，主要分为 4 种——byte（字节型）、short（短整型）、int（整数型）和 long（长整型），如表 2－1－3 所示。

表 2－1－3　整型变量

| 类型名 | 占用存储空间 | 取值范围 |
| --- | --- | --- |
| byte | 8 位（1 个字节） | －27 ~ 27 －1 |
| short | 16 位（2 个字节） | －215 ~ 215 －1 |
| int | 32 位（4 个字节） | －231 ~ 231 －1 |
| long | 64 位（8 个字节） | －263 ~ 263 －1 |

例如：

```
long num == 2200000000L;   //所赋的值超出 int 型的取值范围,后面必须加上字母 L
long num = 198L;        //所赋的值未超出 int 型的取值范围,后面可以加上字母 L
long num = 198;         //所赋的值未超出 int 型的取值范围,后面可以省略字母 L
```

### 2. 浮点型变量

浮点型变量用来存储小数数值，分为两种：float（单精度浮点数型）和 double（双精度浮点数型）。两种浮点型变量所占存储空间的大小以及取值范围如表 2－1－4 所示。

表 2-1-4　浮点型变量

| 类型名 | 占用存储空间 | 取值范围 |
|---|---|---|
| float | 32 位（4 个字节） | 1.4E-45~3.4E+38，-3.4E+38~-1.4E-45 |
| double | 64 位（8 个字节） | 4.9E-324~1.7E+308，-1.7E+308~-4.9E-324 |

例如：

```
float f = 123.4f ;//为一个 float 型变量赋值,后面必须加上字母 f
double d1 = 100.1;//为一个 double 型变量赋值,后面可以省略字母 d
double d2 = 199.3d;//为一个 double 型变量赋值,后面可以加上字母 d
```

在程序中也可以为一个浮点型变量赋一个整数数值，例如：

```
float f = 100;//声明一个 float 型变量并赋整数值
double d = 100;//声明一个 double 型变量并赋整数值
```

### 3. 字符型变量

字符型变量用于存储一个单一字符，一个字符型变量占用 2 个字节。例如：

```
char c = 'a'; //为一个字符型变量赋字符 a
char ch = 97;//为一个字符型变量赋整数 97,相当于赋字符 a
```

### 4. 布尔型变量

布尔型变量用来存储布尔值，布尔型的变量只有两个值：true 和 false。例如：

```
boolean flag = false ; //声明一个布尔型变量,初始值为 false
flag = true ;//改变 flag 变量的值为 true
```

## 三、数据类型转换

Java 的数据类型转换必须遵循以下规则。

（1）boolean（布尔型）不能与其他数据类型进行相互转换。

（2）char（字符型）、byte（字节型）、short（短整型）之间可任意进行相互转换。

（3）数值型数据类型之间由低精度向高精度转换，系统自动完成转换；由高精度向低精度转换，则需要强制转换，如图 2-1-3 所示。

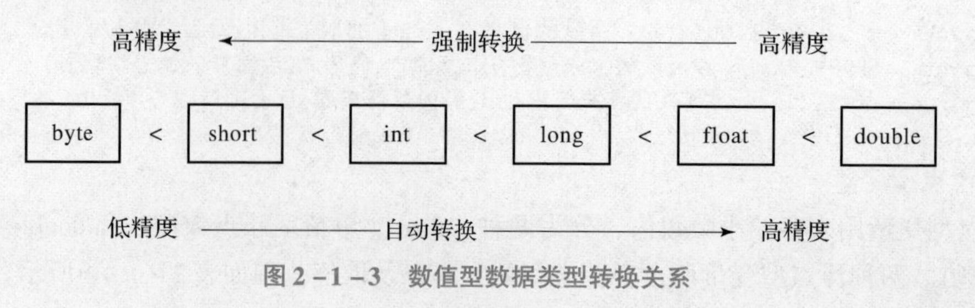


图 2-1-3　数值型数据类型转换关系

①自动转换：也称为隐式类型转换，由系统自动完成。

②强制转换：也称为显式类型转换，强制转换很可能存在精度的损失，因此使用时需要谨慎，其语法格式如下：

```
(目标数据类型)变量名
```

例如：

```
double d = 3.10;//自动转换
int n = (int )d; //强制转换,n =3
```

## 四、变量的作用域

变量需要先定义后使用，但这并不意味着变量定义后一定可以使用。变量需要在它的作用范围内才可以使用，这个作用范围称为变量的作用域。在 Java 程序中，变量一定被定义在某对大括号中，该大括号所包含的代码区域便是这个变量的作用域。

**【例 2-1-1】**变量作用域应用实例。代码如下：

```
1  public class P2_1_1 {
2      public static void main(String[] args) {
3          int x = 12;
4          {
5              int y = 96;
6              {
7                  int z = 0;
8                  y = x;
9              }
10             System. out .println("x is" + x);
11             System. out .println("y is" + y);
12             z = x;
13             System. out .println("z is" + z);
14         }
15         System. out .println("x is" + x);
16     }
17 }
```

运行结果如图 2-1-4 所示。

```
Exception in thread "main" java.lang.Error: Unresolved compilation problems:
        z cannot be resolved to a variable
        z cannot be resolved to a variable

        at class17.P2_1_1.main(P2_1_1.java:12)
```

图 2-1-4　运行结果

【说明】

变量 z 定义在程序第 6 ~9 行的大括号中，引用变量 z 则在第 12 行，超出了它的作用域，因此报错。

**案例**

在计算机中存放并输出个人基本信息表，需要在内存中开辟内存空间，即需要定义变量。定义变量时要明确各数据的类型，学号是整型，姓名是字符串型，性别是字符型，年龄是整型，成绩是浮点型。

代码如下：

```
public class Student {
    public static void main(String[] args) {
        int no = 1;
        String name = "张三";
        char gendar = '男';
        int age = 19;
        float score =90;
        System. out .println("学生信息");
        System. out .println("no" + "name" + "gender" + "age" + "score");
        System. out .println(no + " " + name + " " +gendar +" " +age + "  " +score);
    }
}
```

运行结果如图 2-1-5 所示。

```
学生信息
no name gender age score
1  张三    男     19  90.0
```

图 2-1-5 运行结果

**任务工单**

| 任务名称 | 图书信息管理系统——图书信息数据类型设计 | | | | |
|---|---|---|---|---|---|
| 班级 | | 学号 | | 姓名 | |
| 任务要求 | （1）修改【例 2-1-1】，使程序可以正常运行，并写出运行结果。<br>（2）图书信息包括图书编号、书名、作者、出版社、单价、状态，请合理定义这些变量的数据类型。 | | | | |
| 任务实现 | | | | | |
| 异常及待解决问题记录 | | | | | |

# 2.2 Java 运算符

## 学习目标

(1) 了解 Java 运算符。

(2) 掌握 Java 运算符的优先级及运算规则。

## 相关知识

### 2.2.1 算术运算符

算术运算符是用于处理四则运算的符号，如表 2-2-1 所示。

表 2-2-1 算术运算符

| 运算符 | 运 算 | 范 例 | 结果 |
| --- | --- | --- | --- |
| + | 正号 | +3 | 3 |
| - | 负号 | b=4; -b; | -4 |
| + | 加 | 5+5 | 10 |
| - | 减 | 6-4 | 2 |
| * | 乘 | 3*4 | 12 |
| / | 除 | 5/5 | 1 |
| % | 取模（即算术中的求余数） | 7%5 | 2 |
| ++ | 自增（前） | a=2; b= ++a; | a=3; b=3; |
| ++ | 自增（后） | a=2; b=a++; | a=3; b=2; |
| -- | 自减（前） | a=2; b= -a; | a=1; b=1; |
| -- | 自减（后） | a=2; b=a-; | a=1; b=2; |

【说明】

(1) 在进行自增“++”和自减“--”的运算时，如果运算符 ++ 或 -- 放在操作数的前面则是先进行自增或自减运算，再进行其他运算；反之，如果运算符放在操作数的后面，则是先进行其他运算，再进行自增或自减运算。

(2) 在进行除法运算时，如果除数和被除数都为整数，则得到的结果也是一个整数。如果除法运算有小数参与，则得到的结果是一个小数。例如，2 510/1 000 属于整数之间相除，会忽略小数部分，得到的结果是 2，而 2.5/10 的结果为 0.25。

（3）在进行取模（%）运算时，运算结果的正负取决于被模数（% 左边的数）的符号，与模数（% 右边的数）的符号无关。例如，(-5)%3 = -2，而 5%(-3) = 2。

### 2.2.2　赋值运算符

赋值运算符的作用是将常量、变量或表达式的值赋给某个变量，如表 2-2-2 所示。

表 2-2-2　赋值运算符

| 运算符 | 运 算 | 范 例 | 结 果 |
|---|---|---|---|
| = | 赋值 | a=3; b=2; | a=3; b=2; |
| += | 加等于 | a=3; b=2; a+=b; | a=5; b=2; |
| -= | 减等于 | a=3; b=2; a-=b; | a=1; b=2; |
| *= | 乘等于 | a=3; b=2; a*=b; | a=6; b=2; |
| /= | 除等于 | a=3; b=2; a/=b; | a=1; b=2; |
| %= | 模等于 | a=3; b=2; a%=b; | a=1; b=2; |

【说明】

（1）在赋值过程中，运算顺序为从右往左，将右边表达式的结果赋给左边的变量。

（2）可以通过一条赋值语句对多个变量进行赋值。例如：

```
int x, y, z;
x = y = z = 5;//为 3 个变量同时赋值
```

下面的写法在 Java 中是错误的：

```
int x = y = z = 5;//错误
```

（3）在表 2-2-2 中，除了“=”，其他都是特殊的赋值运算符，以“+=”为例，x += 3 就相当于 x = x + 3，首先会进行加法运算 x+3，再将运算结果赋给变量 x。“-=”“*=”“/=”“%=”运算符都可依此类推。

（4）在为变量赋值时，如果两种类型彼此不兼容，或者目标类型的取值范围小于源类型，则需要进行强制转换，然而在使用“+=”“-=”“*=”“/=”“%=”运算符进行赋值时，强制类型转换会自动完成，程序不需要进行任何显式声明。

### 2.2.3　比较运算符

比较运算符用于对两个数值或变量进行比较，其结果是一个布尔值，如表 2-2-3 所示。

表 2－2－3　比较运算符

| 运算符 | 运 算 | 范 例 | 结 果 |
|---|---|---|---|
| == | 相等 | 4 ==3 | false |
| ! = | 不等于 | 4! =3 | true |
| < | 小于 | 4 <3 | false |
| > | 大于 | 4 >3 | true |
| <= | 小于等于 | 4 <=3 | false |
| >= | 大于等于 | 4 >=3 | true |

## 2.2.4　逻辑运算符

逻辑运算符用于对布尔型数据进行操作，其结果仍是一个布尔型数据，如表 2－2－4 所示。

表 2－2－4　逻辑运算符

| 运算符 | 运 算 | 范 例 | 结 果 |
|---|---|---|---|
| ! | 非 | ! true | false |
| | | ! false | true |
| && | 短路与 | true && true | true |
| | | true && false | false |
| | | false && false | false |
| | | false && true | false |
| ‖ | 短路或 | true ‖ true | true |
| | | true ‖ false | true |
| | | false ‖ false | false |
| | | false ‖ true | true |

【说明】

（1）逻辑运算符可以针对结果为布尔值的表达式进行运算，例如 x >3&&y! = 0。

（2）运算符“&&”表示与操作，当且仅当运算符两边的操作数都为 true 时，其结果才为 true，否则结果为 false。当运算符“&”和“&&”的右边为表达式时，两者在使用上还有一定的区别：在使用“&”进行运算时，无论左边的表达式为 true 还是 false，右边的表达式都会进行运算；在使用“&&”进行运算时，如果左边的表达式为 false，则右边的表达式不会进行运算，因此“&&”被称作短路与。

（3）运算符“‖”表示或操作，当运算符“‖”两边的任何一个操作数的值为 true 时，其结果均为 true；只有当两边的操作数都为 false 时，其结果才为 false。同与操作类似，“‖”表示短路或，当运算符“‖”左边的表达式为 true 时，右边的表达式就不会进行运算。

### 2.2.5　条件运算符

条件运算符（?:）也称为三元运算符，其运算过程为：如果布尔表达式的值为 true，则返回表达式 1 的值；否则返回表达式 2 的值。其语法格式如下：

```
布尔表达式？表达式 1:表达式 2
```

例如：

```
String mark = (score >= 60) ? "及格" : "不及格";
```

### 2.2.6　位运算符

位运算符主要用于对操作数的每个二进制位进行运算，其操作数的类型是整型以及字符型，运算的结果是整型，如表 2 – 2 – 5 所示。

表 2 – 2 – 5　位运算符

| 运算符 | 说明 | 示例 |
|---|---|---|
| <<= | 左移位赋值运算符 | C <<=2 等价于 C = c <<2 |
| >>= | 右移位赋值运算符 | C >>=2 等价于 C = C >>2 |
| & = | 按位与赋值运算符 | C& =2 等价于 C – C&2 |
| ^ = | 按位异或赋值运算符 | c^ =2 等价于 C = C^2 |
| \| = | 按位或赋值运算符 | C \| =2 等价于 C – C \|2 |

### 2.2.7　运算符的优先级

对一些比较复杂的表达式进行运算时，要明确表达式中所有运算符参与运算的先后顺序，通常把这种顺序称作运算符的优先级。接下来通过表 2 – 2 – 6 来展示 Java 中运算符的优先级，最高优先级的运算符在表的最上面，最低优先级的运算符在表的最下面。

表 2 – 2 – 6　运算符的优先级

| 类别 | 运算符 | 关联性 |
|---|---|---|
| 后缀 | ( )、[ ]、.（点运算符） | 从左到右 |
| 一元 | ++ 、 -- 、!、 ~ | 从右到左 |

续表

| 类别 | 运算符 | 关联性 |
| --- | --- | --- |
| 乘性 | *、/、% | 从左到右 |
| 加性 | +、- | 从左到右 |
| 移位 | >>、>>>、<< | 从左到右 |
| 关系 | >>=、<<= | 从左到右 |
| 相等 | ==、!= | 从左到右 |
| 按位与 | & | 从左到右 |
| 按位异或 | ^ | 从左到右 |
| 按位异 | \| | 从左到右 |
| 逻辑与 | && | 从左到右 |
| 逻辑或 | \|\| | 从左到右 |
| 条件 | ?: | 从左到右 |
| 赋值 | =、+=、-=、*=、%=、>>=、<<=、&=、^=、\|= | 从左到右 |
| 逗号 | , | 从左到右 |

**案例**

求下面3个表达式的值。

```
(1)int  test1 ='a'+'b';
(2)char  test2 ='a'+'b';
(3)char  test3 ='a'+1 ;
```

【分析】

（1）test1 变量的返回值为 int 型，赋值号右边为字符'a'的 ASCII 码（97）与字符'b'的 ASCII 码（98）之和。

（2）test2 变量的返回值为 char 型，赋值号右边先求字符'a'的 ASCII 码（97）与字符'b'的 ASCII 码（98）之和 195，再返回 ASCII 码为 195 的字符。

（3）test2 变量的返回值为 char 型，赋值号右边先求字符'a'的 ASCII 码（97）加上 1，得到 98，再返回 ASCII 码为 98 的字符。

运行结果如下：

```
test1 =195,test2 =Ã,test3 =b
```

## 任务工单

<table>
<tr><td>任务名称</td><td colspan="5">小任务</td></tr>
<tr><td>班级</td><td></td><td>学号</td><td></td><td>姓名</td><td></td></tr>
<tr><td>任务要求</td><td colspan="5">1. 填空<br>
<pre>
public class fill1{
    public static void main(String [ ]args){
        int a =10;
        ______ b = a +3.4f +5.6;
        System.out.println("b = " +b);
    }
}
</pre>
2. 输出下图。<br>
<pre>
 !  !  !  !  !  !
   \\-  -//
   (-0 0-)
    \(_)/
     \u/
</pre>
</td></tr>
<tr><td>任务实现</td><td colspan="5"></td></tr>
<tr><td>异常及待解决<br>问题记录</td><td colspan="5"></td></tr>
</table>

# 2.3 流程控制语句

## 学习目标

（1）掌握顺序结构、选择结构、循环结构程序设计的特点。
（2）能运用 if、switch 语句进行选择结构程序设计。
（3）熟练掌握 while、do…while、for 语句的书写格式、执行流程及特点。
（4）正确理解循环嵌套的执行流程。

## 相关知识

### 2.3.1 顺序结构

#### 一、数据输入

使用 java. util 包的 Scanner 类获取数据的 next( )方法系列。
（1）nextInt( )：输入整数。
（2）nextLine( )：输入字符串。
（3）nextDouble( )：输入双精度数。
（4）next( )：输入字符串（以空格作为分隔符）。

**【例 2-3-1】** 代码如下：

```
1  public class P2_3_1 {
2      public static void main(String[] args) {
3      Scanner scanner = new Scanner(System. in); //创建 Scanner 对象
4          System. out .println("请输入一个字符串:");
5          String inputString = scanner.nextLine(); //读取一行文本
6          System. out .println("请输入一个整数:");
7          int inputInt = scanner.nextInt(); //读取整数
8          System. out .println("请输入一个浮点数:");
9          double inputDouble = scanner.nextDouble(); //读取浮点数
10         System. out .println("您输入的字符串是:" + inputString);
11         System. out .println("您输入的整数是:" + inputInt);
12         System. out .println("您输入的浮点数是:" + inputDouble);
13         scanner.close(); //关闭 scanner
14     }
15 }
```

运行结果如图 2-3-1 所示。

```
请输入一个字符串：
hello
请输入一个整数：
50
请输入一个浮点数：
20.8
您输入的字符串是：hello
您输入的整数是：50
您输入的浮点数是：20.8
```

图 2-3-1　运行结果

## 二、数据输出

（1）System. out. println( )：在控制台上打印输出。例如：

```
System.out.println("Hello, World!");
System.out.println(123);
```

（2）System. out. print( )：该方法与 System. out. println( )类似，但它不会在输出后添加新行。例如：

```
System.out.print("Hello, ");
System.out.print("World!");  //输出将会是 "Hello, World!",没有新行
```

（3）格式化输出：Java 的 System. out. printf( )方法允许使用格式化字符串输出数据，这与 C 语言中的 printf( )函数非常相似。例如：

```
System.out.printf("Hello, % s! Your score is % d.", "Alice", 95);
```

### 2.3.2　选择结构

## 一、单分支 if 语句

单分支 if 语句的语法格式如下：

```
  if(表达式){
代码块;
}
```

如图 2-3-2 所示，当表达式的值为 true 时，则执行代码块，否则任何代码块都不执行。

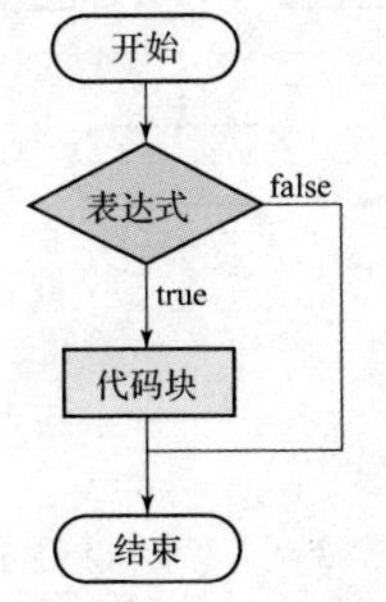


图 2-3-2　单分支 if 语句流程

**【例 2－3－2】** 通过键盘输入年龄，如果年龄不小于 18 岁，则在屏幕上显示“我是成年人，可以独立实施民事法律行为!”。代码如下：

```
1  public class P2_3_2 {
2      public static void main(String[] args) {
3          Scanner s = new Scanner(System. in);
4          System. out .println("请输入你的年龄:");
5          int age = s.nextInt();
6          if (age > = 18) {
7              System. out .println("我是成年人,可以独立实施民事法律行为!");
8          }
9      }
10 }
```

运行结果如图 2－3－3 所示。

```
请输入你的年龄:
20
我是成年人，可以独立实施民事法律行为!
```

图 2－3－3　运行结果

## 二、双分支 if…else 语句

双分支 if…else 语句的语法格式如下：

```
   if(表达式){
  代码块1;
}else{
  代码块2;
}
```

如图 2－3－4 所示，当表达式的值为 true 时，执行代码块 1，否则执行代码块 2。

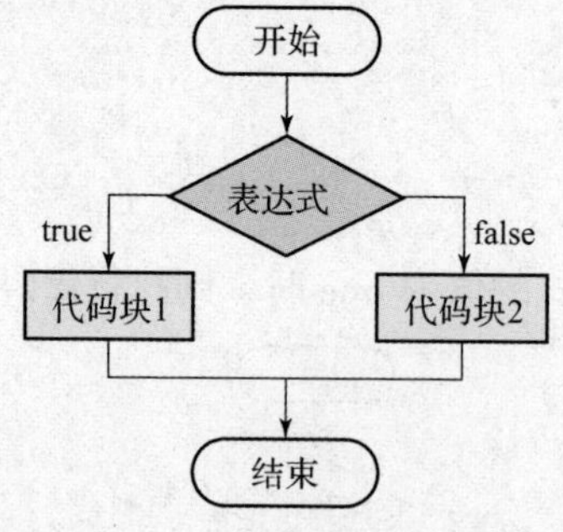


图 2－3－4　双分支 if…else 语句流程

**小贴士**

不要误认为 if…else 语句是两条语句（if 语句和 else 语句）。它们都属于同一个 if 语句。else 子句不能作为语句单独使用，它必须与 if 配对使用。

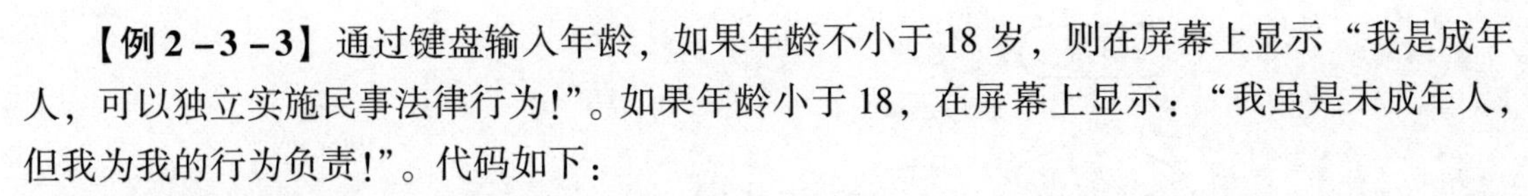

【例 2-3-3】通过键盘输入年龄，如果年龄不小于 18 岁，则在屏幕上显示“我是成年人，可以独立实施民事法律行为!”。如果年龄小于 18，在屏幕上显示：“我虽是未成年人，但我为我的行为负责!”。代码如下：

```
1  public class P2_3_3 {
2      public static void main(String[] args) {
3          Scanner s = new Scanner(System.in);
4          System.out.println("请输入你的年龄:");
5          int age = s.nextInt();
6          if (age < 0) {
7              return;
8          }
9          if (age >= 18) {
10             System.out.println("我是成年人,可以独立实施民事法律行为!");
11         } else {
12             System.out.println("我虽是未成年人,但我为我的行为负责!");
13         }
14     }
15 }
```

运行结果如图 2-3-5 所示。

```
请输入你的年龄:
10
我虽是未成年人，但我为我的行为负责!
```

图 2-3-5　运行结果

## 三、多分支 if…else if…else 语句

多分支 if…else if…else 语句的语法格式如下：

```
    if(表达式1){
    代码块1;
}else if(表达式2){
    代码块2;
}
    …
}else if(表达式n){}
    代码块n;
}else{
    代码块n+1;
}
```

如图 2-3-6 所示，根据不同的表达式值确定执行哪条语句，语句中测试条件的顺序为表达式 1、表达式 2、……，一旦遇到表达式的值为 true，就执行该条件下的代码块。若所有表达式的值均为 false，且语句中选用了 else 部分，则执行代码块 n，否则不执行任何代码块。

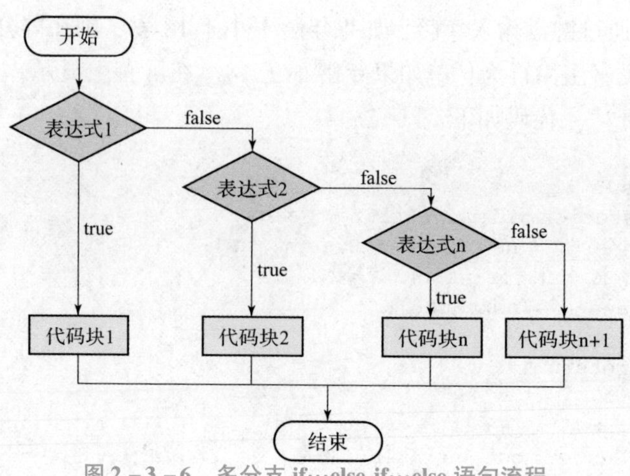


图 2-3-6　多分支 if…else if…else 语句流程

**小贴士**

最后一个 else 后面不要再写条件。

【例 2-3-4】按照屏幕上的提示信息输入 1~12，选择四季中的一种。代码如下：

```
1  public class P2_3_3 {
2      public static void main(String[] args) {
3          Scanner scanner = new Scanner(System.in);
4          System.out.print("请输入一个月份(1~12):");
5          int month = scanner.nextInt();
6          if (month == 3 || month == 4 || month == 5) {
7              System.out.println("春季");
8          } else if (month == 6 || month == 7 || month == 8) {
9              System.out.println("夏季");
10         } else if (month == 9 || month == 10 || month == 11) {
11             System.out.println("秋季");
12         } else if (month == 12 || month == 1 || month == 2) {
13             System.out.println("冬季");
14         } else {
15             System.out.println("输入的月份有误");
16         }
17     }
18 }
```

运行结果如图 2-3-7 所示。

```
请输入一个月份(1~12)：6
夏季
```

```
请输入一个月份(1~12)：15
输入的月份有误
```

图 2-3-7　运行结果

## 四、if 语句的嵌套

在 if 语句中又包含一个或多个 if 语句，称为 if 语句的嵌套。使用时应注意 if 与 else 的配对，else 总是与它上面最近的未配对的 if 配对。if 语句的嵌套的语法格式如下：

```
if( )
    if()
        代码块 1;
    else          }内嵌 if
        代码块 2;
else
    if( )
        代码块 3;
    else          }内嵌 if
        代码块 4;
```

## 五、switch 语句

switch 语句是一种常用的多分支选择语句，与 if 语句不同，它只能针对某个表达式的值作出判断，从而决定程序执行哪一段代码块。switch 语句的语法格式如下：

```
  switch(表达式){
  case 常量表达式 1: 代码块 1; [break;]
  case 常量表达式 2: 代码块 2;[break;]
  …
  case 常量表达式 n: 代码块 n; [break;]
  default : 代码块 n +1;
}
```

如图 2-3-8 所示，计算表达式的值并逐个与其后的常量表达式的值比较，当表达式的值与某个常量表达式的值相等时，执行其后的代码块，break 语句是可选的，但如果没有 break 语句，程序会继续执行下一条 case 语句，直到遇到 break 语句。如果表达式的值与所有 case 后的常量表达式均不相等，则执行 default 后的语句。

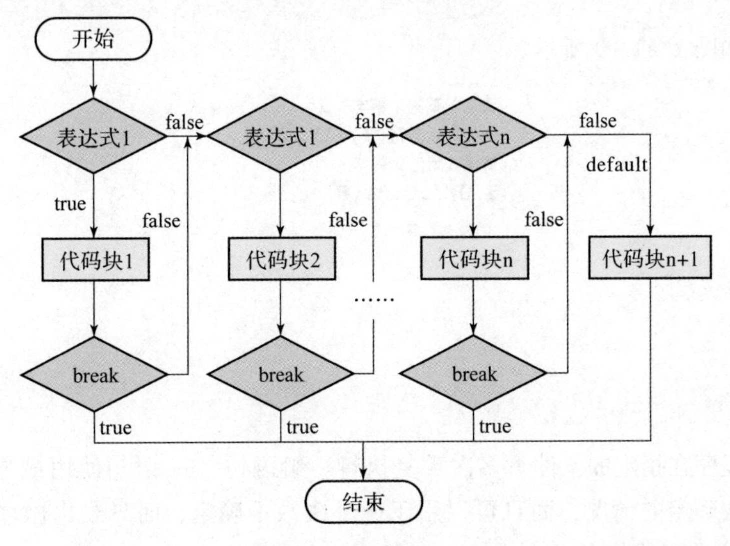


图 2-3-8　switch 语句流程

在使用 switch 语句时还应注意以下几点。

（1）表达式的类型可以是字符型或整型。

（2）case 后的各常量表达式的值不能相同，否则会出现错误。

（3）在 case 后允许有多个语句，可以不用大括号括起来。

（4）各 case 和 default 子句的先后顺序可以变动，而不会影响程序执行结果。

（5）default 子句可以省略。

（6）多个 case 可以共用一组执行语句。

**【例 2-3-5】**输入运算数和四则运算符，输出计算结果。代码如下：

```
1  public class P2_3_4 {
2      public static void main(String[] args) {
3          Scanner sc = new Scanner(System.in);
4          double num1, num2, result;
5          char operator;
6          System.out.print("请输入第一个数字:");
7          num1 = sc.nextDouble();
8          System.out.print("请输入运算符(+,-,*,/):");
9          operator = sc.next().charAt(0);
10         System.out.print("请输入第二个数字:");
11         num2 = sc.nextDouble();
12         switch (operator) {
13         case '+':result = num1 + num2;break;
14         case '-':result = num1 - num2;break;
15         case '*':result = num1 * num2;break;
16         case '/':result = num1 /num2;break;
17         default :System.out.println("无效的运算符!");return;
18         }
19         System.out.println(num1 + "" + operator + "" + num2 + "=" + result);
20     }
21 }
```

运行结果如图 2-3-9 所示。

```
请输入第一个数字：3
请输入运算符(+,-,*,/):*
请输入第二个数字：7
3.0*7.0=21.0
```

图 2-3-9　运行结果

## 2.3.3　循环结构

### 一、for 语句

循环是指程序在指定的条件下多次重复执行一组语句。for 语句使用最为灵活，不仅可以用于循环次数确定的情况，而且可以用于循环次数不确定，而只给出循环结束条件的情况。for 语句的语法格式如下：

```
for(表达式 1;表达式 2;表达式 3) {
   语句;
}
```

for 语句的执行流程如下。

(1) 先求解表达式 1，若其值为 true（值为非 0），则执行 for 语句中指定的代码块，然后执行第（2）步。若其值为 false（值为 0），则结束循环，转到第（3）步。

(2) 转回第（1）步继续执行。

(3) 循环结束，执行 for 语句下面的一条语句。

for 语句流程如图 2-3-10 所示。

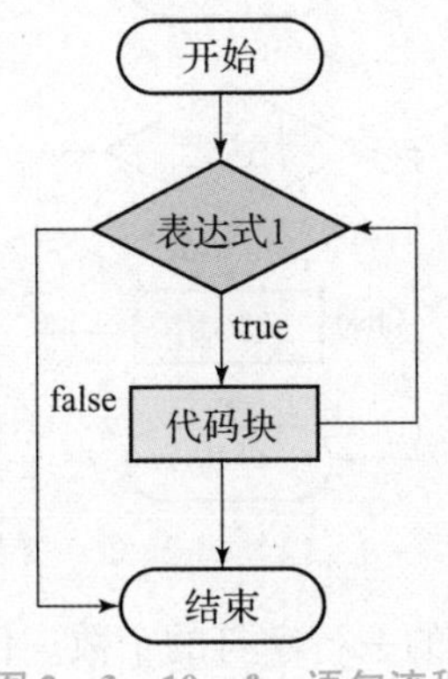


图 2-3-10　for 语句流程

**【例 2-3-6】** 计算 n!。代码如下：

```
1  public class P2_3_5 {
2      public static void main(String[ ] args) {
3          System.out.printf("输入任意一个数:");
4          Scanner input = new Scanner(System.in);//用于接收输入数据
5          int n = input.nextInt();//将输入的数据赋给 n
6          int a = 1;//用于存储阶乘的值
7          for (int i = 1; i < = n; i ++) {
8              a * = i;//等同于 a = a * i >,阶乘运算公式
9              System.out.printf("% d\n", a);//每计算一次就打印一次
10         }
11         System.out.printf("d 的阶乘为:% d\n", a);//打印最后计算的结果
12     }
13 }
```

运行结果如图 2-3-11 所示。

```
输入任意一个数：3
1
2
6
d的阶乘为：6
```

图 2-3-11　运行结果

## 二、while 语句

while 语句会反复进行条件判断，只要表达式的值为 true，则大括号中的代码块就会一直执行，直到表达式的值为 false，while 循环才会结束。while 语句的语法格式如下：

```
while (表达式) {
    代码块;
}
```

while 语句流程如图 2-3-12 所示。

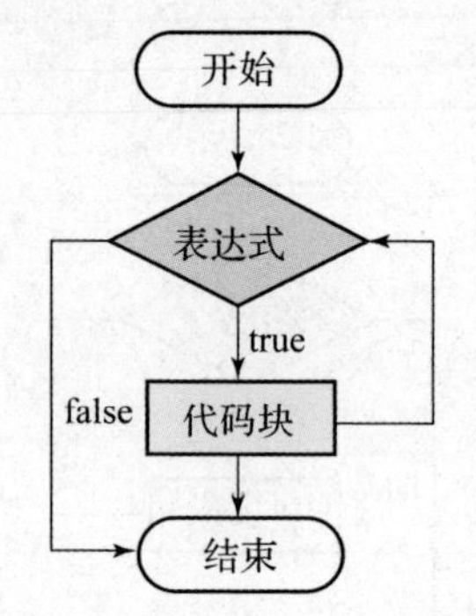


图 2-3-12　while 语句流程

**【例 2-3-7】**统计从键盘输入的一行字符的个数。代码如下：

```
1  public class P2_3_6 {
2      public static void main(String[] args){
3          Scanner sc = new Scanner(System.in);
4          String str;
5          str = sc.nextLine();
6          int chars,num,space,other; //分别表示字母、数字、空格和其他字符的个数
7          int ch;//字符串中的一个字符
8          chars = num = space = other = 0;
9          ch =0;//统计各类字符的个数
10          int len =str.length();
11          while (ch < len){
12              if (str.charAt(ch) > = 'a'&&str.charAt(ch) < = 'z' ||
13                        str.charAt(ch) > = 'A'&&str.charAt(ch) < = 'Z')
14                 chars ++;
15              else if (str.charAt(ch) > = '0'&&str.charAt(ch) < = '9')
16                 num ++;
17              else if (str.charAt(ch) == '')
18                   space ++;
19               else other ++;
20               ch ++;
21              }
22          System.out.printf("字母%d数字%d空格%d他%d\n",chars,num,space,other);
23          }
24  }
```

运行结果如图 2－3－13 所示。

```
12ww  33
字母2数字3空格2他0
```

图 2－3－13　运行结果

## 三、do…while 语句

do…while 语句与 while 语句功能类似，区别是 while 语句需要先判断循环条件，再根据判断结果决定是否执行大括号中的代码块；而 do…while 语句需要先执行一次大括号中的代码块再判断循环条件。do…while 语句的语法格式如下：

```
do {
    代码块;
} while(表达式);
```

do…while 语句流程如图 2－3－14 所示。

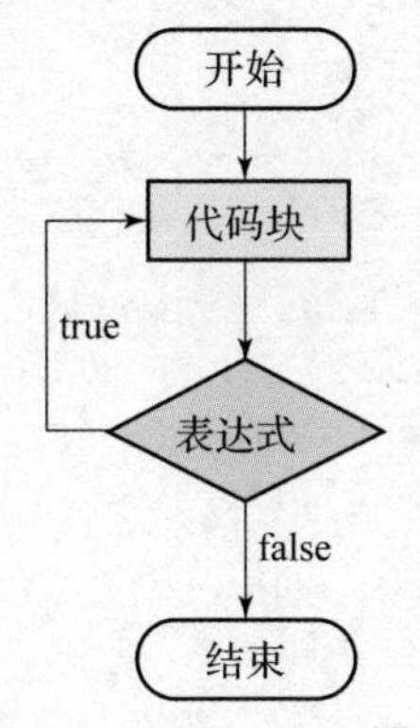


图 2－3－14　do…while 语句流程

**【例 2－3－8】** 求 1～100 的累加和。代码如下：

```
1  public class P2_3_7 {
2      public static void main(String[] args) {
3          int n = 1;
4          int sum = 0;
5          do {
6              sum = sum + n;
7              n++;
8          } while (n <= 10);
9          System.out.println("sum = " + sum);
10     }
11 }
```

运行结果如图 2－3－15 所示。

```
sum=55
```

图 2－3－15　运行结果

## 四、增强 for 语句

Java 5 引入了一种主要用于数组的增强 for 语句，其语法格式如下：

```
  for(声明语句 : 表达式) {
    代码块;
}
```

（1）声明语句：声明新的局部变量，该变量的类型必须与数组元素的类型匹配。其作用域限定在循环语句块内，其值与此时数组元素的值相等。

（2）表达式：要访问的数组名，或者返回值为数组的方法。

**【例 2-3-9】** 增强 for 语句应用实例。代码如下：

```
1  public class P2_3_8 {
2      public static void main(String[] args) {
3          int [] numbers = { 10, 20, 30, 40, 50 };
4          for (int x : numbers) {
5              System. out .print(x);
6              System. out .print(",");
7          }
8          System. out .print("\n");
9          String[] names = { "James", "Larry", "Tom", "Lacy" };
10         for (String name : names) {
11             System. out .print(name);
12             System. out .print(",");
13         }
14     }
15 }
```

运行结果如图 2-3-16 所示。

```
10,20,30,40,50,
James,Larry,Tom,Lacy,
```

图 2-3-16　运行结果

## 五、跳转语句

### 1. break 语句

当 break 语句出现在 switch 语句中时，其作用是终止某个 case 语句并跳出 switch 结构；当它出现在循环语句中时，其作用是跳出当前循环语句，执行后面的代码块，如图 2-3-17 所示。当 break 语句出现在嵌套循环的内层循环中时，它只能跳出内层循环。

### 2. continue 语句

在循环语句中，如果希望立即终止本次循环，并执行下一次循环，就需要使用 continue 语句，如图 2-3-18 所示。

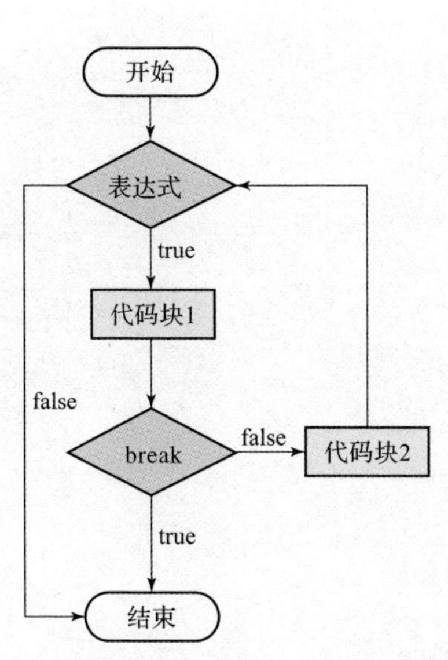

图 2-3-17　break 语句流程

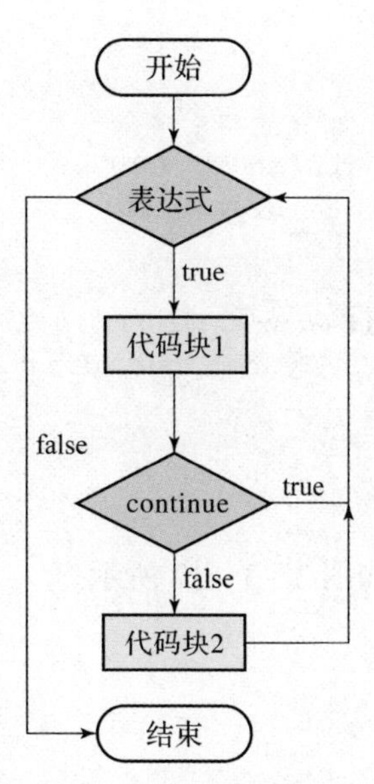

图 2-3-18　continue 语句流程

3. return 语句

return 语句作为一个无条件的分支，无须判断条件即可发生。return 语句主要有两个用途：一是表示一个方法返回的值（假定返回值类型不是 void），二是导致该方法退出并返回值。根据方法的定义，每个方法都有返回类型，该类型可以是基本类型，也可以是对象类型，同时每个方法都必须有一个结束标志，return 语句起到了这个作用。在返回类型为 void 的方法中有一个隐含的 return 语句，因此，在这类方法中 return 可以省略不写。

【例 2-3-10】设计一个猜数字游戏。计算机随机产生一个 0～100 的数，提供 5 次机会，每次通过键盘输入数字，计算机可以给出"大了""小了""挑战成功"3 种提示。当出现"挑战成功"时，游戏结束，如果 5 次都猜不中，则计算机给出正确结果。代码如下：

```
3  public class P2_3_9 {
4      public static void main(String[] args) {
5          Scanner scanner = new Scanner(System.in);
6          Random random = new Random();
7          int i = random.nextInt(101);//生成 0~100 的随机整数,作为答案
8          int count = 5;
9          while (count > 0) {
10             System.out.println("请输入你猜想的数字");
11             int guess = scanner.nextInt();
12             if (i == guess) {
13                System.out.println("挑战成功," + "答案为" + i);
14                break;
15             }
16             if (i < guess) {
17                System.out.println("大了");
18             }
19             if (i > guess) {
20                System.out.print("小了" + "\t");
```

```
21            }
22            count --;
23            if (count > 1) {
24                System. out .println("你还有" + count + "次机会");
25            }
26          }
27          if (count == 0) {
28              System. out .println("挑战失败:" + "答案为" + i);
29          }
30      }
31  }
```

运行结果如图 2 -3 -19 所示。

```
请输入你猜想的数字
4
小了		你还有4次机会
请输入你猜想的数字
7
小了		你还有3次机会
请输入你猜想的数字
50
大了
你还有2次机会
请输入你猜想的数字
30
挑战成功，答案为30
```

图 2 -3 -19　运行结果

## 六、循环的嵌套

一个循环体内又包含另一个完整的循环结构称为循环的嵌套。内嵌的循环体中还可以嵌套循环，这就是多层循环。

**【例 2 -3 -11】** 打印由“ * ”组成的倒直角三角形，共 5 行，第一行有 5 个“ * ”，每行减少 1 个“ * ”，第 5 行有 1 个“ * ”。代码如下：

```
1  public class P2_3_10 {
2      public static void main(String[] args) {
3          for (int x = 0; x < 5; x ++) {
4              for (int y = x; y < 5; y ++) {
5                  System. out .print(" * ");
6              }
7              System. out .println();
8          }
9      }
10 }
```

运行结果如图 2 -3 -20 所示。

```
*****
****
***
**
*
```

图 2 -3 -20　运行结果

**案例**

实现客户管理系统的用户注册功能。当用户输入正确的用户名时，让用户输入密码，用户有 3 次输入密码的机会，输入密码正确即可进入系统，如果用户 3 次输入密码错误，则无权进入系统。

【分析】

(1) 用户名输入：输入语句。

(2) 3 次密码输入：循环语句。

(3) 3 次密码验证：循环语句。

代码如下：

```
public class P2_18 {
    public static void main(String[ ] args) {
        System. out .println("新用户注册 XXX");
        Scanner scanner = new Scanner(System. in );
        System. out .println("请输入注册的账号");
        String username = scanner.next();
        System. out .println("请输入注册的密码");
        StringBuffer password = new StringBuffer(scanner.next());
        System. out .println("确认密码");
        StringBuffer password2 = new StringBuffer(scanner.next());
        if (password.toString().equals(password2.toString())) {
            System. out .println("两次密码输入一致");
        } else {
            System. out .println("两次密码输入不一致,请重新输入");
            int p2_length = password2.length();
            password2.delete(0, p2_length);
            password2 = new StringBuffer(scanner.next());
            password2.toString().equals(password.toString());
        }
        for (int i = 1; i < = 3; i ++) {
             System. out .println("请输入您的用户名");
             Scanner sc = new Scanner(System. in );
             String s1 = sc.next();
             System. out .println("请输入您的密码");
             String s2 = sc.next();
             if (username.equals(s1) && password.toString().equals(s2)) {
                 System. out .println("登录成功");
                 break;
             } else if (i < 3) {
                 System. out .println("您还有" + (3 - i) + "次机会");
             } else {
                 System. out .println("由于您 3 次输入错误,强制退出");
             }
        }
    }
}
```

## 任务工单

| 任务名称 | 管理系统——用户登录功能 | | | | |
|---|---|---|---|---|---|
| 班级 | | 学号 | | 姓名 | |
| 任务要求 | 实现管理系统的用户登录功能。用户有 3 次输入用户名和密码的机会，只有用户名和密码均正确方可登录系统，否则给出错误提示并退出系统。 | | | | |
| 任务实现 | | | | | |
| 异常及待解决问题记录 | | | | | |

## 2.4 数组

### 学习目标

（1）了解数组的概念。
（2）掌握一维数组的定义、初始化和引用。
（3）掌握二维数组的定义、初始化和引用。
（4）掌握数组排序的算法。

### 相关知识

数组属于复合数据类型，是由若干类型相同的元素组成的一个有序的数据集合。数组中各元素的类型相同，而且各元素有序排列，所有元素共用一个名称。数组中的元素可以通过下标访问。

### 2.4.1 一维数组

#### 一、一维数组的声明

一维数组的声明的语法格式如下：

```
数组元素类型 数组名[]；  或    数组元素类型[] 数组名；
```

例如：

```
char s[];              //s 的每个元素都是 char 类型的
String[]  points;      //points 的每个元素都是 String 类型的
Car Cars[];            //Cars 的每个元素都是类 Car 类型的
```

#### 二、一维数组的初始化

声明数组仅是为这个数组指定数组名和数组元素的类型，并不为数组元素分配实际的存储空间。数组经过初始化后，其长度就会固定，不再改变。几种常见的一维数组的初始化方法如下。

（1）在声明的同时初始化，例如：

```
int[] array = {1, 2, 3, 4, 5};
```

（2）先声明，后分别赋值，例如：

```
int[] array = new int[3];
array[0] = 1; array[1] = 2; array[2] = 3; array[3] = 4; array[4] = 5;
```

（3）动态初始化，只指定数组长度，元素默认初始化值（整型默认为0），例如：

```
int [] array = new int [5];
```

（4）使用 for 循环初始化，例如：

```
int [] array = new int [5];
for (int i = 0; i < array.length; i ++) {
    array[i] = i + 1;
}
```

（5）使用 Arrays. fill( )方法初始化，例如：

```
int [] array = new int [5];
Arrays.fill(array, 10); //所有元素都会被填充为 10
```

（6）匿名内部类初始化，例如：

```
int [] array = new int [] {1, 2, 3, 4, 5};
```

## 三、一维数组元素的引用

**数组初始化后可通过数组名与下标来引用数组中的元素。一维数组元素的引用的语法格式如下：**

```
数组名[数组下标];
```

**注意：在 Java 中，数组下标是指元素在数组中的位置，数组下标从 0 开始，最大值为数组长度 -1。例如：**

```
int [] a =new int [10],n =3;
a[3] =25;          //正确,下标是整数常量
a[3 +6] =90;       //正确,下标是整数表达式
a[n] =45;          //正确,下标是整型变量
a[10] =8;          //错误,下标是 10,越界
```

**【例 2 -4 -1】** 用数组求解斐波那契（Fibonacci）数列（1，1，2，3，5，8，13，21，…）的前 20 项。代码如下：

```
1  public class P2_4_1 {
2      public static void main(String[] args) {
3          int i;
4          int fib[] = new int [20];
5          fib[0] = 1;
6          fib[1] = 1;
7          for (i = 2; i < 20; i ++) {
```

```
8              fib[i] = fib[i - 2] + fib[i - 1];
9          }
10         for(i = 0; i < 20; i++) {
11             if(i% 5 == 0)
12                 System.out.println("\n");
13             System.out.print("\t" + fib[i]);
14         }
15     }
16 }
```

运行结果如图 2 - 4 - 1 所示。

```
1      1      2      3      5
8      13     21     34     55
89     144    233    377    610
987    1597   2584   4181   6765
```

图 2 - 4 - 1　运行结果

## 2.4.2　二维数组

### 一、二维数组的声明

二维数组的声明的语法格式如下：

```
类型说明符 数组名[][];   或   类型说明符[][] 数组名;
```

例如，创建图 2 - 4 - 2 所示二维数组，其每行都有 4 个元素，称为矩阵数组。代码如下：

```
int a[][] = new int[2][];
a[0] = new int[4];
a[1] = new int[4];
```

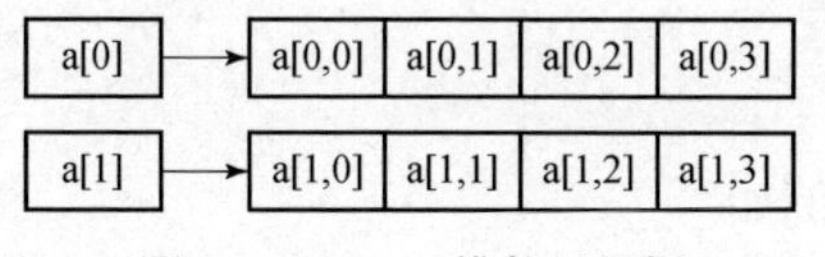


图 2 - 4 - 2　二维数组示例

在 Java 中还可以创建非矩阵数组，其声明的语法格式如下：

```
类型 数组名[][] = new 类型[length1][length2];
```

例如：

```
int a[][] = new int[4][];
a[0] = new int[2];     //第 1 行声明为 2 个元素
a[1] = new int[4];     //第 2 行声明为 4 个元素
a[2] = new int[6];     //第 3 行声明为 6 个元素
a[3] = new int[8];     //第 4 行声明为 8 个元素
```

以下声明方式是错误的：

```
int errarr1[2][3];//不允许静态声明数组
int errarr2[][] = new int [][4]; //数组的维数说明顺序应从高维到低维
int errarr3[][4] = new int [3][4];//数组维数的指定只能出现在new运算符之后
```

## 二、二维数组的初始化

定义好数组之后，需要对数组进行初始化。常用的二维数组初始化方法如下。

（1）直接指定数组元素，这种方式必须指定数组的行数和列数。例如：

```
int [][] array1 = {{1, 2, 3},{4, 5, 6},{7, 8, 9}};
```

（2）先指定行数，然后对每行分别初始化。例如：

```
int [][] array2 = new int [3][3];
array2[0] = new int [] {1, 2, 3};
array2[1] = new int [] {4, 5, 6};
array2[2] = new int [] {7, 8, 9};
```

（3）先指定行数，然后对每行分别初始化，不需要指定列数。例如：

```
int [][] array3 = new int [3][];
array3[0] = new int [] {1, 2, 3};
array3[1] = new int [] {4, 5, 6};
array3[2] = new int [] {7, 8, 9};
```

（4）先指定行数和列数，然后逐个赋值。例如：

```
int [][] array4 = new int [2][2];
array4[0][0] = 1;
array4[0][1] = 2;
array4[1][0] = 4;
array4[1][1] = 5;
```

（5）使用循环初始化。例如：

```
int [][] array5 = new int [3][3];
for (int i = 0; i < array5.length; i ++) {
    for (int j = 0; j < array5[i].length; j ++) {
        array5[i][j] = (i + 1) * (j + 1);
    }
}
```

## 三、二维数组元素的引用

二维数组元素的引用的语法格式如下：

```
数组名[下标1][下标2];
```

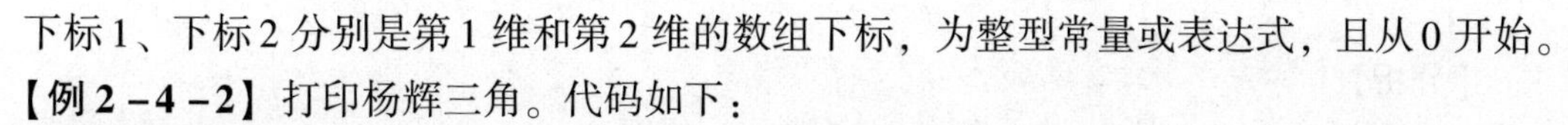


下标1、下标2分别是第1维和第2维的数组下标，为整型常量或表达式，且从0开始。

【例2-4-2】打印杨辉三角。代码如下：

```
1  public class P2_4_2 {
2      public static void main(String[] args) {
3          int [][] arr = new int [10][];
4          //变量二维数组,动态给每个元素赋值
5          for (int i = 0; i < arr.length; i ++) {
6              arr[i] = new int [i + 1];
7              //给每个元素赋值
8          for (int j = 0; j < arr[i].length; j ++) {
9              if (j == 0 || j == arr[i].length - 1) {
10               arr[i][j] = 1;
11             } else {
12               arr[i][j] = arr[i - 1][j] + arr[i - 1][j - 1];
13             }
14           }
15       }
16       for (int i = 0; i < arr.length; i ++) {
17           for (int j = 0; j < arr[i].length; j ++) {
18               System. out .print(arr[i][j] + "\t");
19           }
20           System. out .println();
21       }
22     }
23 }
```

运行结果如图2-4-3所示。

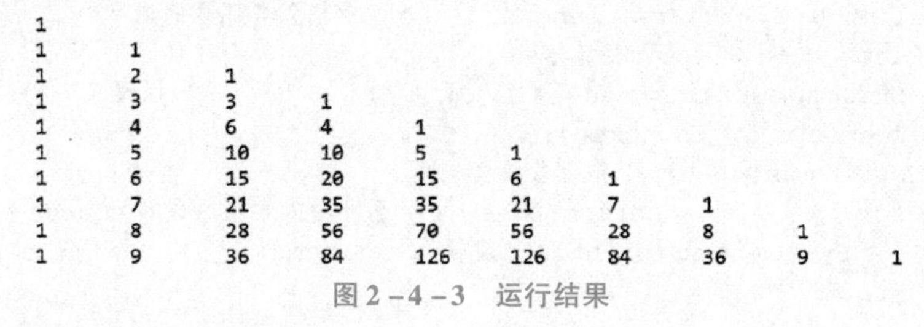

图2-4-3 运行结果

**案例**

课程小组成员的成绩记录如表2-4-1所示，计算课程小组成员的总分和平均分，并按总分从高到低排序。

表2-4-1 成绩记录

| NO | NAME | GENDER | AGE | LESSON1 | LESSON2 | LSESSON3 | LESSON4 | SUM | AVG |
|---|---|---|---|---|---|---|---|---|---|
| 1 | 张三 | 男 | 18 | 88 | 85 | 90 | 95 | | |
| 2 | 李四 | 男 | 19 | 90 | 95 | 90 | 90 | | |
| 3 | 王五 | 女 | 18 | 70 | 75 | 60 | 65 | | |
| 4 | 赵六 | 女 | 18 | 85 | 90 | 95 | 95 | | |
| 5 | 钱七 | 男 | 19 | 75 | 80 | 85 | 80 | | |

【分析】

(1) 定义一维数组存放学生姓名、课程名、总分、平均分。

(2) 定义二维数组存放学生各课程成绩。

代码如下:

```
public class P2_23 {
    public static void main(String[] args) {
        Scanner in = new Scanner(System. in );
        System. out .println("输入学生人数:");
        int person = in.nextInt();
        System. out .println("输入课程数目:");
        int coursenum = in.nextInt();
        String[] name = new String[person];
        String[] course = new String[coursenum];
        int [][] aa = new int [person][coursenum];
        int [] sum = new int [person];//定义总分数组 int
        int [] avg = new int [person];//定义平均分数组
        String[] ss = new String[person]; //整合一行的信息
        int i, j;
        //循环存储课程名称
        for (i = 0; i < name.length; i ++) {//输入各学生的姓名
            System. out .println("请输入第" + (i + 1) + "的学生姓名");
            name[i] = in.next();//接受输入的学生姓名
        }
        //循环存储学生各课程成绩
        for (i = 0; i < course.length; i ++) {//输入各课程的名字
            int s = 0;//定义 s 存储分数
            System. out .println("请输入第" + (i + 1) + "的课程名字");
            course[i] = in.next();
            String ss1 = "";//整合各课程成绩
            for (j = 0; j < person; j ++) {//循环输入学生的成绩 course
                System. out .println("请输入" + name[j] + "的" + course[i] + "
                成绩:");
                aa[i][j] = in.nextInt();
                s + = aa[i][j];//将循环的分数累加到 s
                ss1 + = aa[i][j] + "\t";
                sum[i] = s;
                avg[i] = s /coursenum;
                //计算平均分
                ss[i] = name[i] + "\t" + ss1 + sum[i] + "\t" + avg[i];
            }
        }
    //根据总分,整合后的名次转整行互换
    for (i = 0; i < sum.length - 1; i ++) {//冒泡排序输出名次
        for (j = 0; j < sum.length - 1; j ++) {
            if (sum[j] < sum[j + 1]) {
                int t = sum[j];
```

```
                String t2 = ss[j];
                sum[j] = sum[j + 1];
                ss[j] = ss[j + 1];
                sum[j + 1] = t;
                ss[j + 1] = t2;
            }
        }
    }
    for (i = 0; i < course.length; i ++) {//输出效果
        System.out.print("\t" + course[i]);
    }
    System.out.println("\t总分\t平均分\t名次");//打印总分列
    for (i = 0; i < person; i ++) {
       System.out.print(name[i]);
         for (j = 0; j < coursenum; j ++) {
             System.out.print("\t" + aa[i][j]);
    }
    System.out.print("\t" + sum[i] +"\t" + avg[i]);//输出总分
    System.out.println( "\t第" + (i + 1) + "名");
    }
  }
}
```

## 任务工单

<table>
<tr><td>任务名称</td><td colspan="5">成绩统计</td></tr>
<tr><td>班级</td><td></td><td>学号</td><td></td><td>姓名</td><td></td></tr>
<tr><td>任务要求</td><td colspan="5">（1）批量统计课程小组各成员所修课程的最高分和最低分。<br>（2）按总分从高到低排序。</td></tr>
<tr><td>任务实现</td><td colspan="5"></td></tr>
<tr><td>异常及待解决问题记录</td><td colspan="5"></td></tr>
</table>

**练习题**

**1. 选择题**

（1）在下面的赋值语句中，会产生编译错误的是（　　）。

A. float a =2.0;　　　　B. double b =2.0;

C. int c =2;　　　　D. long d =2;

（2）下列语句执行后，i 的值是（　　）。

```
int i =8,j =16;
if(i -1 >j)i --; else j --;
```

A. 15　　B. 16　　C. 7　　D. 8

（3）下列语句执行后，k 的值是（　　）。

```
int i =10,j =18,k =30;
switch( j -i)
{
    case 8:k ++;
    case 9:k + =2;
    case 10:k + =3;
    default :k/=j;
}
```

A. 31　　B. 32　　C. 2　　D. 33

（4）下列语句执行后，i 的值是（　　）。

```
for( int i =0,j =1;j <5;j + =3)i =i +j;
```

A. 4　　B. 5　　C. 6　　D. 7

（5）设有定义“float x =3.5f，y =4.6f，z =5.7f;”，则以下表达式中，值为 true 的是（　　）。

A. x >y || x >z　　　　B. x! =y

C. z >(y +x)　　　　D. x <y&!(x <z)

（6）以下由 for 语句构成的循环执行的次数是（　　）。

```
for( int i =0; true;i ++);
```

A. 有语法错误，不能执行　　　　B. 无限次

C. 执行 1 次　　　　D. 一次也不执行

（7）下列语句执行后，k 的值是（　　）。

```
int m =3,n =6,k =0;
while((m ++) <( --n)) ++k;
```

A. 0　　B. 1　　C. 2　　D. 3

（8）下面的初始化语句中错误的是（　　）。

A. char str[] ="hello";　　　　B. char str[100] ="hello";

C. char str[] ={'h','e',,'','o'};　　　　D. char str[] ={'hello'};

（9）定义了 int 型一维数组 a[10]后，下面的引用中错误的是（　　）。

A. a[0] =1;　　　　B. a[10] =2;

C. a[0] =5 * 2;　　　　　　　　D. a[1] = a[2] * a[0];

(10) 下列二维数组初始化语句中，正确的是（　　）。

A. float b[2][2] = {0.1,0.2,0.3,0.4};

B. int a[][] = {{1,2},{3,4}};

C. int a[2][] = {{1,2},{3,4}};

D. float a[2][2] = {0};

2. 填空题

(1) 表达式(int)((double)(3)/2)的值是＿＿＿＿＿＿。

(2) 表达式 5.3 + (int)(8.5 + 4.6)/3%4 的值是＿＿＿＿。

(3) 执行语句“inta,b,c;a = 1;b = 3;c = (a + b > 3? ++a:b ++);”后，b 的值为＿＿＿＿＿＿。

(4) 已知字符“工”的 Unicode 值为 49，则执行语句“System. out. println('1' + 2);”输出＿＿＿＿＿＿。

(5) 数组的最小下标是＿＿＿＿＿＿。

3. 编程题

(1) 求 1 + (1 + 2) + (1 + 2 + 3) + … + (1 + 2 + 3 + 4 + … + 10)的值。

(2) 使用 while（for）语句在屏幕上输出题图 2 - 1 所示图形。

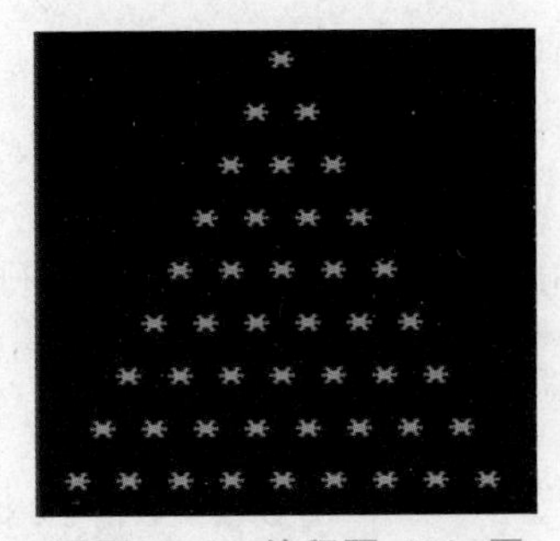

题图 2 - 1　编程题（2）图

第3章

# 面向对象（一）

——不积跬步，无以至千里

# 3.1 面向对象封装性

## 学习目标

（1）理解面向对象程序设计思想及其重要性。
（2）理解类与对象的概念。
（3）理解面向对象封装性。
（4）掌握类、对象的创建与使用方法。

## 相关知识

### 3.1.1 面向对象程序设计思想

#### 一、面向对象程序设计

面向对象是一种符合人类思维习惯的编程思想。现实生活中存在各种各样的事物，这些事物有自己的属性和行为，它们之间存在千丝万缕的联系。在程序中使用对象映射现实中的事物，使用属于对象的函数（方法）实现事物的行为，使用对象的关系描述事物之间的联系。为了实现事物的行为，专门为对象开发了实现功能的函数，别的对象是无法使用该函数，这就是面向对象程序设计。当然，一个应用程序包含多个对象，通过多个对象的相互配合来实现应用程序的功能，这样当应用程序的功能发生变动时，只需要修改个别对象即可，从而使代码更容易维护。

例如设计歌咏比赛选手计分系统，要求去除10位评委打分中的最高分、最低分，求出平均分作为选手最终得分。以此为例进行面向对象程序设计。

（1）事物（对象）：评委、分数、选手。

（2）行为（函数）。

①评委（Judges）：评分（inputScore()函数）。

②分数（Scores）：求总分（sumScore()函数）、求平均分（avgScore()函数）、求最高分（maxScore()函数）、求最低分（minScore()函数）、求最终得分（lastScore()函数）。

③选手（Player）：查询得分情况（queryScore()函数）。

（3）行为实施（函数调用）。

①评委提交分数：

```
新建一个评委对象 Judges j1 =new Judge();
调用评委对象特有函数评分 j1. inputScore();
```

②查询选手所有得分：

```
新建一个选手对象 Player p1 =new Player();
调用选手对象特有函数查询得分 p1. queryScore();
```

③计算最终得分：

```
新建一个分数对象 Scores s1 =new Scores ();
调用分数对象特有函数求最终得分 s1. lastScore ();
```

每个实现特定功能的函数都只有指定对象才能使用，其他对象无法使用。例如，选手对象 p1 调用评委录入成绩的方法 inputScore( )，系统报无此方法的错误。这种专门为某个对象设计程序，就是面向对象程序设计思想。

## 二、面向对象的特性

### 1. 封装性

封装性是面向对象的核心思想，将对象的属性和行为封装起来，不需要外界知道具体实现细节，这就是封装思想。Java 通过类把属性和方法封装起来，减少程序间的依赖，降低程序的复杂性，提高程序的安全性。

### 2. 继承性

继承性主要描述类与类之间的关系。通过继承，可以在无须重新编写原有类的情况下，对原有类的功能进行扩展。继承不仅提高了代码的复用性，提高了开发效率，还为程序的维护提供了便利。

### 3. 多态性

多态性是指在程序中允许出现重名现象，即在一个类中定义的属性和方法被其他类继承后，它们可以具有不同的数据类型或表现出不同的行为，这使同一个属性和方法在不同的类中具有不同的语义。多态性使程序具有良好的可扩展性，易于维护。

## 3.1.2　类与对象

面向对象程序设计，简单地理解就是对某个对象进行程序开发，以描述现实中事物的状态和行为。Java 中的对象是由类派生的，先有类后有对象。

## 一、类

类是对某类事物的抽象描述。可以把类理解成模型，模型的作用是为创建事物提供依据。例如鞋模、橡皮泥玩具模型等，用同一个鞋模制作的鞋都一样；用同一个橡皮泥玩具模型印出来的玩具形状也都一样；同理，用同一个类创建的对象的结构也是一样的。

在类中可以定义成员变量和成员方法，成员变量用于描述对象的特征，也被称作属性；成员方法用于描述对象的行为，可简称为方法。类的语法格式如下：

```
  class  类名{
  //成员变量
  //成员方法
}
```

**【例 3-1-1】** 人类被抽象为 Person 类，人类的姓名、性别、年龄等特征被抽象为

name、sex、age 三个成员变量，人类打招呼的行为被抽象为成员方法 sayHi( )。代码如下：

```
1  public class Person {
2      //成员变量
3      public String name; //姓名
4      public String sex; //性别
5      public int age; //年龄
6      //成员方法
7      public void sayHi(){System.out.println("Hi");}
8  }
```

## 二、对象

类是一类事物的抽象，对象是类的具体化。如果说类是模型，那么根据模型制作的事物就是对象，它们具有相同的特征和行为。例如，根据鞋模制作的鞋就是一个对象，使用橡皮泥玩具模型印出来的一个玩具形状也是一个对象。因此，类只有一个，对象可以有无数个；类是抽象的，对象是具体的。

### 1. 创建对象

创建对象的语法格式如下：

```
类名 对象名 =new 类名();
```

例如，根据 Person 类实例化一个人类对象，代码如下：

```
Person  p1 =new Person();
```

（1）Person p1：声明一个对象 p1，是 Person 数据类型。

（2）new Person( )：调用构造方法实例化对象 p1，此时如图 3－1－1 所示，在内存中为对象 p1 的每个成员变量都分配了相应的内存单元，每个对象的每个成员变量的存储都是独立的。与此同时，Java 虚拟机会自动对成员变量进行初始化，针对不同类型的成员变量，Java 虚拟机会赋予其不同的初始值，如表 3－1－1 所示。

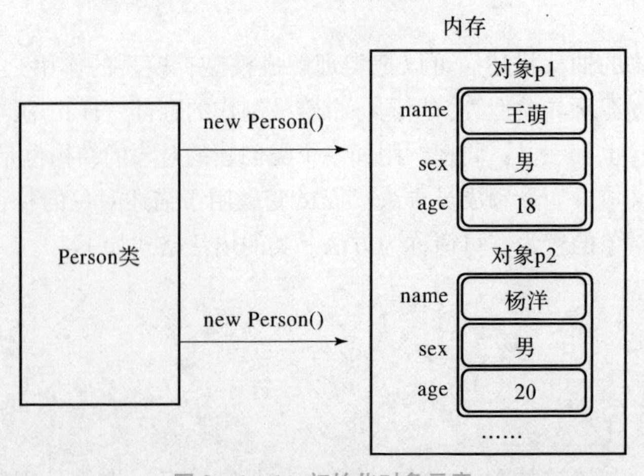


图 3－1－1 初始化对象示意

表3-1-1　成员变量初始化

| 成员变量类型 | 初始值 | 成员变量类型 | 初始值 |
|---|---|---|---|
| byte | 0 | double | 0.0D |
| short | 0 | char | 空字符，'\u0000' |
| int | 0 | boolean | false |
| long | 0L | 引用数据类型 | null |
| float | 0.0F | — | — |

### 2. 使用对象

创建对象后，可以通过对象名访问类中的所有成员变量和成员方法，其语法格式如下：

```
对象名.成员变量
对象名.成员方法
```

**【例3-1-2】**代码如下：

```
1  public class Person {
2      String name;
3      String sex;
4      int age;
5      public void sayHi(){System.out.println("Hi");}
6      //主函数,演示对象的引用
7      public static void main(String[] args) {
8          Person p1 = new Person();//创建第1个对象
9          p1.name = "王萌";//为第1个对象的name变量赋值
10         P1.age = 18
11         Person p2 = new Person();//创建第2个对象
12         p2.name = "杨洋";//为第2个对象的name变量赋值
13         p1.sayHi();//调用对象的sayHi()方法
14         p2.sayHi();
15         System.out.println("我的姓名是:" + p1.name + ",我的年龄是:" + p1.age);
16         System.out.println("我的姓名是:" + p2.name + ",我的年龄是:" + p2.age);
17     }
18 }
```

运行结果如图3-1-2所示。

```
Hi
Hi
我的姓名是：王萌，我的年龄是：18
我的姓名是：杨洋，我的年龄是：0
```

图3-1-2　运行结果

【说明】

p1、p2分别为Person类的两个实例对象，两次输出的name值不同，说明p1和p2对象

拥有各自的 name 属性，修改 p1 对象的 name 属性不会影响 p2 对象的 name 属性。

## 三、类的设计

在 Java 中，对象是通过类创建的。因此，在进行程序设计时，最重要的就是类的设计。下面以学生选课系统为例，介绍如何从需求中抽取类、设计类。

### 1. 抽取类

学生选课系统主要处理哪位学生选修哪位教师的哪门课程，并记录选修成绩。学生选课系统涉及的主要事物有学生、教师、课程、成绩。因此，需要设计 4 个类，分别为学生类（记录学生信息）、教师类（记录教师信息）、课程类（记录课程信息）、成绩类（记录学生选修成绩信息）。

### 2. 设计类

（1）学生类（Student）：学号、姓名、性别、年龄。

（2）教师类（Teacher）：姓名、性别、年龄、职称。

（3）课程类（Course）：课程号、课程名、学时、学分、教师姓名。

（4）成绩类（Grade）：学号、课程号、成绩。

代码如下：

```
1   public class Student {
2       String sid, name, sex;
3       int age;
4       void introduce() {
5           System.out.println("I am a student,my name is:" + name);
6       }
7   }
8   class Teacher{
9     String tid, name, sex;
10    int age;
11      void myWork() {
12          System.out.println("I am a teacher,my name is:" + name);
13      }
14  }
15  class Course{
16      String cid,cname ;
17      int classhours,credit;
18      void introduce() {
19          System.out.println("课程名::" + cname + ",学分:" + credit);
20      }
21  }
22  class Grade{
23      String sid, cid;
24      double score;
25      void outputScore() {
26          System.out.println(sid + "同学" + cid + "课程的成绩:" + score);
27      }
28  }
```

### 3.1.3　类的封装性

在设计类时，成员变量不允许外界随意访问，这就需要实现类的封装。所谓类的封装，是指在定义一个类时，将类中的属性私有化，即使用 private 关键字修饰。私有属性只能在它所在的类中被访问，如果外界想要访问私有属性，则需要提供一些使用 public 修饰的公有方法，其中包括用于获取属性值的访问方法（getXxx()）和设置属性值的设置方法（setXxx()），具体如下：

```
1  public class Person {
2      //成员变量私有化
3      private String name,sex;
4      private int age;
5      //访问、设置方法
6      public String getName() {
7         return name;
8      }
9      public void setName(String name) {
10         this.name = name;
11     }
12     public String getSex() {
13         return sex;
14     }
15     public void setSex(String sex) {
16         if (sex.equals("男") || sex.equals("女"))
17             this.sex = sex;
18     else
19             System.out.println("非法数据");
20     }
21     public int getAge() {
22         return age;
23     }
24     public void setAge(int age) {
25         if (age >=0)
26             this.age = age;
27         else
28             System.out.println("非法数据");
29     }
30     //成员方法
31     public void sayHi(){System.out.println("Hi");}
32 }
```

测试类如下：

```
3  public class Test {
4      public static void main(String[] args) {
5          Person p1 = new Person();
6          p1.setName("王萌");
```

```
7          System. out .println("第 1 次修改姓名:" +p1.getName());
8          p1.name = "张三";
9         System. out .println("第 1 次修改姓名:" +name);
10      }
11    }
```

运行结果如图 3-1-3 所示。

```
Exception in thread "main" java.lang.Error: Unresolved compilation problems:
	The field Person.name is not visible
	name cannot be resolved to a variable

	at chapter03.Test.main(Test.java:9)
```

图 3-1-3　运行结果

【说明】

Person 类的所有成员变量都私有化，为所有成员变量提供了公有访问方法（getXxx()）、设置方法（setXxx()），并对性别、年龄参数进行非法数据过滤。在测试类中，使用成员变量只能通过访问方法、设置方法，否则系统将给出错误提示。

**案 例**

设计小型花店管理系统，实现对花店中鲜花、顾客及鲜花销售的管理。

【分析】

（1）鲜花。

静态特征：花名、价格、库存量；动态行为：库存查询。

（2）顾客。

静态特征：顾客的姓名、性别、年龄；动态行为：说话、买东西。

（3）销售单。

静态特征：销售单号、顾客姓名、销售日期；动态行为：销售记录查询。

### 1. 类的设计

1）Flower

成员变量：fname（String）、fprice（double）、fcount（int）。

成员方法：int query()。

2）Customer

成员变量：cname（String）、csex（String）、cage（int）。

成员方法：void talk()、void buy()。

3）Sale

成员变量：sid（String）、cname（String）、saledate（Date）。

成员方法：void query()。

### 2. 类的定义

代码如下：

```
public class Flower {
    private String fname;
    private String flanguage;
    private int fcount;
    public String getFname() {
        return fname;
    }
    public void setFname(String fname) {
        this.fname = fname;
    }
    public double getFprice() {
        return fprice;
    }
    public void setFprice(double fprice) {
        this.fprice = fprice;
    }
    public int getFcount() {
        return fcount;
    }
    public void setFcount(int fcount) {
        this.fcount = fcount;
    }
    //成员方法
    public int query() {
        return fcount;
    }
}
//客户类Customer
public class Customer {
    private String cname;
    private String csex;
    private int cage;
    public String getCname() {
        return cname;
    }
    public void setCname(String cname) {
        this.cname = cname;
    }
    public String getCsex() {
        return csex;
    }
    public void setCsex(String csex) {
        this.csex = csex;
    }
    public int getCage() {
        return cage;
    }
    public void setCage(int cage) {
        this.cage = cage;
```

```
    }
    //成员方法
    public void talk() {
        System. out .println("HI");
    }
    public void buy() {
        System. out .println("买花喽!!!");
    }
}
//销售单类
public class Sale {
    private String sid,cname;
    private Date Saledate;
    public String getSid() {
        return sid;
    }
    public void setSid(String sid) {
        this .sid = sid;
    }
    public String getCname() {
        return cname;
    }
    public void setCname(String cname) {
        this .cname = cname;
    }
    public Date getSaledate() {
        return Saledate;
    }
    //成员方法
    public void setSaledate(Date saledate) {
        Saledate = saledate;
    }
    public void query() {
        System. out .println( "Sale [sid =" + sid + ", cname =" + cname + ", Saledate ="
                + Saledate + "]");
    }
}
```

任务工单

| 任务名称 | 图书信息管理系统——图书类定义（Book） | | | | |
|---|---|---|---|---|---|
| 班　级 | | 学号 | | 姓名 | |
| 任务要求 | （1）图书类信息包含图书编号（bookID）、书名（bookName）、作者（writer）、出版社（press）、单价（price）、状态（status）。定义 Book 类，合理使用数据类型。<br>（2）为 Book 类添加 QueryBook( )成员方法，该方法要求根据图书编号，输出该编号的书名，合理设置参数及返回值类型。<br>（3）创建对象，并测试 QueryBook( )成员方法。 | | | | |
| 任务实现 | | | | | |
| 异常及待解决问题记录 | | | | | |

# 3.2 方法与 static 关键字

## 学习目标

（1）掌握构造方法的定义与使用方法。
（2）掌握方法的重载。
（3）了解垃圾回收机制。
（4）掌握 static 关键字的使用方法。

## 相关知识

### 3.2.1 成员方法

#### 一、成员方法的设计

在类中定义成员变量和成员方法。成员变量用于描述对象的特征，成员方法用于描述对象的行为，可简称为方法。方法由 4 部分组成，分别为方法名、参数列表、返回值数据类型、方法体，4 个部分缺一不可。方法的语法格式如下：

```
返回值数据类型  方法名([参数列表]){
  //方法体
}
```

（1）方法名采用“驼峰”式命名。
（2）允许有多个参数，参数间用逗号隔开，无论是否有参数，均保留小括号。
（3）如果有返回值，则方法体中必须 return 返回值，且数据类型必须一致；如果没有返回值，则返回值数据类型设置为 void。

#### 二、成员方法的调用

调用成员方法的语法格式如下：

```
对象名．方法名();
```

**【例 3-2-1】**以加减运算为例，介绍方法的定义与调用。代码如下：

```
1 public class Calculate {
2      private double a,b;
3      public double getA() {return a;}
4      public void setA(double a) {this.a = a;}
5      public double getB() {return b;}
6      public void setB(double b) {this.b = b;}
7      //成员方法
```

```
8      public void setValue(double p1,double p2) {
9          a = p1;
10         b = p2;
11     }
12     public double add() {//返回计算结果
13         return a + b;
14     }
15     public double sub() {//输出计算结果
16         System.out.println(a - b);
17     }
18     public static void main(String[] args) {
19         Calculate01 cal = new Calculate01();
20         cal.setValue(2, 3); //调用 setValue()方法,对两个成员变量赋值
21         System.out.println(cal.add());//调用 add()方法,并输出结果
22         cal.sub();//调用 sub()方法,并输出结果
23     }
24 }
```

运行结果如图 3 - 2 - 1 所示。

```
5.0
-1.0
6.0
0.6666666666666666
```

图 3 - 2 - 1 运行结果

### 3.2.2 构造方法

类中定义的方法如果同时满足以下 3 个条件，则该方法称为构造方法。

(1) 方法名与类名相同。

(2) 在方法名的前面没有返回值数据类型的声明。

(3) 在方法中不能使用 return 语句返回一个值，但可以单独写 return 语句作为方法的结束。

【例 3 - 2 - 2】以 Person 类为例，定义构造方法。代码如下：

```
1  public class Person {
2      private String name , sex;
3      private int age; //无参构造方法
4      public Person() {
5          System.out.println("我是无参构造方法");
6      }
7      //带参构造方法
8      public Person(String n) {
9          name = n;
10     }
11     public static void main(String[] args) {
12         Person p = new Person(); //调用无参构造方法
13         Person p = new Person("王萌"); //调用带参构造方法
14         System.out.println("我的名字是:" + p.name);
15     }
```

```
16  }
```

运行结果如图 3－2－2 所示。

我的名字是：王萌

图 3－2－2　运行结果

【说明】

如果在类中没有定义构造方法，则系统为该类提供一个默认的无参构造方法；一旦在类中定义了构造方法，则系统不再提供默认的无参构造方法。

### 3.2.3　方法的重载

在一个类中定义多个同名方法，每个同名方法的参数类型或参数数量不同，即构成方法的重载。系统根据用户提供的参数调用这些同名方法。

**【例 3－2－3】** 代码如下：

```
1  public class Person {
2      //成员变量
3      private String name ,sex;
4      private int age;
5      //构造方法重载(方法名相同,参数类型或参数数量不同)
6      public Person(String n) {
7          name = n;
8      }
9      public Person(int a) {
10         age = a;
11     }
12     public Person(String n, String s, int a) {
13         name = n;
14         sex = s;
15         age = a;
16     }
17     //成员方法重载(方法名相同,参数类型或参数数量不同)
18     public void introduce1(String n,  int a) {
19         name = n;
20         age = a;
21         System.out.println("我的名字是:" +name + ",年龄:" +age);
22     }
23     public void introduce2() {
24         System.out.println("我的名字是:" +name + ",性别:" +sex + ",年龄:" +age);
25     }
26     public static void main(String[] args) {
27         Person p1 =new Person("王萌");//调用带 1 个 String 型参数的构造方法
28         p1.introduce2();
29         Person p2 =new Person(10);//调用带 1 个整型参数的构造方法
30         p2.introduce2();
31         Person p3 =new Person("王萌","男",20);//调用带 3 个参数的构造方法
32         p3.introduce2();//调用无参的 introduce2()方法
33         p3.introduce1("李明红", 30);//调用带 2 个参数的 introduce1()方法
34     }
35 }
```

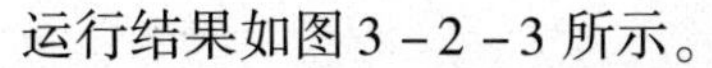

运行结果如图 3-2-3 所示。

```
我的名字是：王萌，性别：null，年龄：0
我的名字是：null，性别：null，年龄：10
我的名字是：王萌，性别：男，年龄：20
我的名字是：李明红，年龄：30
```

图 3-2-3　运行结果

### 3.2.4　static 关键字

在 Java 中，static 关键字可以修饰成员变量、成员方法以及代码块等，被 static 关键字修饰的成员具备一些特殊性，下面对这些特殊性进行讲解。

#### 一、静态变量

对于通过 new 关键字创建的实例对象，系统将为它们分配各自的内存空间，但在实际程序开发过程中，如果希望一个类的所有实例对象共享一个内存空间的数据，则可通过 static 关键字修饰成员变量来实现，该变量被称作静态变量。静态变量具有如下特性。

（1）静态变量可以通过“对象名 . 静态变量名”的形式访问，也可以通过“类名 . 静态变量名”的形式访问。

（2）static 关键字只能用于修饰成员变量，不能用于修饰局部变量。

**【例 3-2-4】** 为 Person 类添加静态变量 static String Country，用来描述人的国籍。静态变量内存空间分配示意如图 3-2-4 所示。

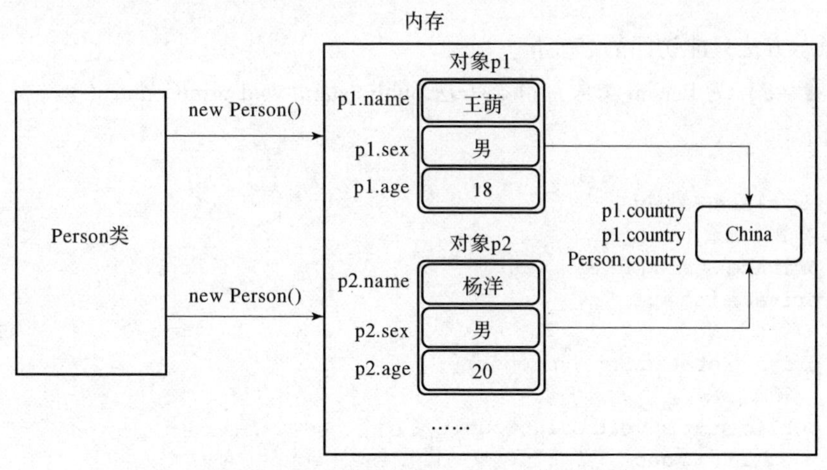


图 3-2-4　静态变量内存空间分配示意

代码如下：

```
1  public class Person {
2      //成员变量
3      private String name , sex;
4      private int age;
5      //静态变量
6      private static String country;
```

```
7       public static void main(String[] args) {
8           Person p1 = new Person();
9           Person p2 = new Person();
10          p1.country = "China";
11          System.out.println("p1 的国籍是:" + p1.country);
12          System.out.println("p2 的国籍是:" + p2.country);
13          System.out.println("p2 的国籍是:" + Person.country);
14      }
15  }
```

运行结果如图 3-2-5 所示。

```
p1的国籍是：China
p2的国籍是：China
p2的国籍是：China
```

图 3-2-5　运行结果

## 二、静态方法

在程序开发过程中，有时需要在不创建对象的情况下调用某个方法，也就是使该方法不必和对象绑定在一起。要实现这样的效果，只需要在成员方法前加上 static 关键字即可，通常这种方法称为静态方法。静态方法具有如下特性。

（1）静态方法可以通过“对象名．方法名”的形式访问，也可以通过“类名．方法名”的形式访问。

（2）静态方法只能访问静态变量。

**【例 3-2-5】** 为 Person 类添加静态方法 public static void printCountry(){}，用来输出国籍。代码如下：

```
1   public class Person {
2       //成员变量
3       private String name, sex;
4       private int age;
5       //静态变量
6       private static String country;
7       //静态方法
8       public static void printCountry() {
9           System.out.println("我的国籍是:" + country);
10          System.out.println("我的姓名是:" + name);//报错,因为 name 不是静态变量
11      }
12      public static void main(String[] args) {
13          Person p1 = new Person();
14          p1.country = "China";
15          p1.printMessage();
16          Person.printMessage();
17      }
18  }
```

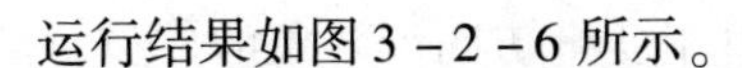

运行结果如图3-2-6所示。

```
我的国籍是：China
我的国籍是：China
```

图3-2-6　运行结果

### 三、静态代码块

在Java类中，使用大括号包围的若干行代码称为一个代码块，用static关键字修饰的代码块称为静态代码块。当类被加载时，静态代码块会执行，由于类只加载一次，所以静态代码块只执行一次。因此，在Java程序中，通常使用静态代码块对类的成员变量进行初始化。

**【例3-2-6】**代码如下：

```
1  public class StaticExample {
2     //静态代码块
3     static {
4         System.out.println("测试类的静态代码块执行了");
5     }
6     public static void main(String[] args) {
7         Person01 p1 = new Person01();
8         Person01 p2 = new Person01();
9     }
10 }
11 class Person01 {
12     static {
13         System.out.println("Person类中的静态代码块执行了");
14     }
15 }
```

运行结果如图3-2-7所示。

```
测试类的静态代码块执行了
Person类中的静态代码块执行了
```

图3-2-7　运行结果

【说明】

从运行结果可以看出，程序中的两段静态代码块都执行了。Java虚拟机首先加载类StaticExample，同时执行该类的静态代码块，接着会执行main()方法。main()方法中创建了两个Person对象，但在两次实例化对象的过程中，静态代码块中的内容只输出了一次，这说明静态代码块在类第一次使用时才会被加载。

## 3.2.5　垃圾对象与垃圾回收

### 一、垃圾对象

（1）在Java中，null是一种特殊的常量，表示该变量不指向任何一个对象。在程序运行过程中，如果某个对象被赋值为null，则表示该对象会失去引用，成为垃圾对象。

（2）当对象被实例化后，在程序中可以通过对象的引用变量访问该对象的成员。需要注意的是，当没有任何变量引用这个对象时，它自动成为垃圾对象，不能再被使用。

## 二、垃圾回收

在 Java 中，当一个对象成为垃圾对象后仍会占用内存空间，针对这种情况，Java 引入了垃圾回收机制，程序员不必处理垃圾回收问题，Java 虚拟机会自动回收垃圾对象所占用的内存空间。Java 虚拟机在两种情况下会启动垃圾回收器：①垃圾对象堆积到一定程度；②调用 System. gc( ) 方法来通知 Java 虚拟机立即进行垃圾回收。

当一个对象在内存中被释放时，它的 finalize( ) 方法会被自动调用，因以在类中通过定义 finalize( ) 方法来观察对象何时被释放。下面通过一个案例演示 Java 虚拟机进行垃圾回收的过程。代码如下：

```
1  public class Person {
2      private String name , sex;
3      private int age;
4      public void sayHi(){System. out .println("Hi");}
5      public void finalize() {
6          System. out .println("垃圾对象被回收了!!!");
7      }
8      public static void main(String[] args) {
9          Person p1 = new Person();
10         Person p2 = new Person();
11         p1 = null ; //成为垃圾对象
12         p2 = null ; //成为垃圾对象
13         System.gc();//通知 Java 虚拟机进行垃圾回收
14     }
15 }
```

运行结果如图 3-2-8 所示。

```
垃圾对象被回收了！！！
垃圾对象被回收了！！！
```

图 3-2-8　运行结果

**案例**

设计新生报到管理模块，需要完成新生信息的确认、宿舍的分配、各学院报到人数统计等功能。录入新生信息，相应学院报到人数要增加 1，如果新生需要住宿，则为其分配宿舍，否则宿舍放空。假定宿舍仅有 100 个，分配宿舍前先判断剩余数量，如果有剩余则分配宿舍，否则提示“暂无宿舍分配”。

### 1. 案例分析

1）新生类

（1）成员变量：学号、姓名、性别、年龄、学院、宿舍。

（2）成员方法：各学院报到人数是该学院每位新生共享的一个数据，应该设置为静态变量，因此采用静态数组存放各学院报到人数；在构造方法中实现人数自动增加1的功能。

（3）宿舍分配方式：①在构造方法中设置宿舍；②调用宿舍分配方法。

2）宿舍类

（1）静态特征：宿舍编号、所在楼栋。

（2）剩余宿舍数是所有宿舍类对象共享的一个数据，因此应该设置为静态变量。

3）统计类

统计类用于统计各学院已报到人数。

**2. 类的定义**

1）学生类

代码如下：

```
class Student{
    private String sid, name, sex; //学号、姓名、性别
    private int age; //年龄
    private String dormId; //宿舍编号
    private String institute; //所在学院
    //静态数组,一个数组元素存储一个学院报到人数
    private static int number[] = new int [5];
    public static int [] getNumber() {
        return number;
    }
    public static void setNumber(int [] number) {
        Student.number = number;
    }
    //分配宿舍方法
    public void setDorm(Dormitory d) {
        dormId = d.getDormId();
    }
    //直接分配宿舍的构造方法
    public Student ( String sid, String name, String sex, int age, String dormId,
String institute) {
    this.sid = sid;
    this.name = name;
    this.sex = sex;
    this.age = age;
    this.dormId = dormId;
    this.institute = institute;
    switch (institute) {
        case "外语":number[0] ++;break ;
        case "信息":number[1] ++;break ;
        case "土建":number[2] ++;break ;
        case "电子":number[3] ++;break ;
        case "经管":number[4] ++;break ;
        }
    }
```

```
    //成员方法
    public void printStu() {
        System.out.println("Student [sid = " + sid + ", name = " + name + ", sex
= " + sex + ", age = " + age + ", dormId = " + dormId
                + ", institute = " + institute + "]");
    }
}
```

2）宿舍类

代码如下：

```
class Dormitory{
    private String dormId, building;
    private static int limit =100; //静态变量
    //get()、set()方法(略)
    public Dormitory(String dormId, String building) {
        if (limit >0) {
            this.dormId = dormId;
            this.building = building;
            limit --;
        }
    }
}
```

3）统计类

代码如下：

```
public class Enrolment {
    public static void count() {//统计各学院入学情况,定义成静态,可直接调用
        int number[] = Student.getNumber();
        System.out.println("        学院                    入学人数");
        System.out.println("       外语学院                     " +number[0]);
        System.out.println("       信息学院                     " +number[1]);
        System.out.println("       土建学院                     " +number[2]);
        System.out.println("       电子学院                     " +number[3]);
        System.out.println("       经管学院                     " +number[4]);
    }
}
```

4）测试类

代码如下：

```
public class Test{
    public static void main(String[] args) {
        Student stu =null ;
        System.out.println("欢迎入学");
        System.out.println("是否住宿(住宿按1,否则按0):");
```

```
        Scanner sc = new Scanner(System. in );
        int n = sc.nextInt();
        if (n ==1) {
            Dormitory dorm = new Dormitory("西101", "小图楼");
            if (dorm.getDormId()! =null ){
            stu = new Student("001", "林丽娟", "女", 19, dorm.getDormId(), "外语");
            }else {
                stu = new Student("001", "林丽娟", "女", 19, "外语");
                System. out .println("已报到,但暂无宿舍分配");
            }
        }
        stu.printStu();
        System. out .println("欢迎入学");
        System. out .println("是否住宿(住宿按1,否则按0):");
        n = sc.nextInt();
        if (n ==1) {
            Dormitory dorm = new Dormitory("西102", "小图楼");
            if (dorm.getDormId()! =null ){
            stu = new Student("002", "王鹏", "男", 19, dorm.getDormId(), "土建");
            }else {
                stu = new Student("002", "王鹏", "男", 19, "土建");
                System. out .println("已报到,但暂无宿舍分配");
            }
        }
        stu.printStu();
        Enrolment.count();
    }
}
```

运行结果如图 3 -2 -9 所示。

```
欢迎入学
是否住宿（住宿按1，否则按0）：
1
Student [sid=001, name=林丽娟, sex=女, age=19, dormId=西101, institute=外语]
欢迎入学
是否住宿（住宿按1，否则按0）：
1
Student [sid=002, name=王鹏, sex=男, age=19, dormId=西102, institute=土建]
    学院            入学人数
外语学院              1
信息学院              0
土建学院              1
电子学院              0
经管学院              0
```

图 3 -2 -9　运行结果

## 任务工单

<table>
<tr><td>任务名称</td><td colspan="5">图书信息管理系统——完善图书类的定义（Book）</td></tr>
<tr><td>班级</td><td></td><td>学号</td><td></td><td>姓名</td><td></td></tr>
<tr><td>任务要求</td><td colspan="5">（1）为 Book 类添加一个图书管理员的静态变量 manager。<br>（2）为 Book 类添加访问/设置方法、构造方法、toString( )方法。<br>（3）创建 Book 类实例，测试 Book 类中的每种方法。</td></tr>
<tr><td>任务实现</td><td colspan="5">（任务 1：记录一个成员变量的访问/设置方法，记录 0 ~ 1 个形参的构造方法即可；任务 2：完整记录）</td></tr>
<tr><td>异常及待解决问题记录</td><td colspan="5"></td></tr>
</table>

## 练习题

**1. 选择题**

（1）类的定义必须包含在（　　）之间。

A. 多重方括号　B. 大括号　C. 双引号　D. 圆括号

（2）下列正确的类的声明是（　　）。

A. public void HH {……}　B. public class Move() {……}

C. public class void number {……}　D. public class Car {……}

**2. 填空题**

（1）面向对象的3个特性是________、________、________。

（2）在Java中，可以用________关键字创建类的实例对象。

（3）定义在类中的变量称为____________，定义在方法中的变量称为____________。

（4）面向对象程序设计的重点是________的设计，________是用来创建对象的模板。

**3. 设计题**

1）JiSuan类

（1）定义一个JiSuan类（计算类）。

（2）为JiSuan类定义double型的a、b两个成员变量。

（3）定义double jia()、double jian()、double cheng()、double chu()方法（加、减、乘、除方法）。

（4）声明一个JiSuan对象，测试4个方法。

2）KFC类

（1）定义一个KFC类。

（2）为KFC类定义String型的address、tel两个成员变量。

（3）定义肯德基门店的介绍方法String introduce()，该方法返回当前肯德基门店地址和电话。

（4）创建一个具体的肯德基门店，调用String introduce()方法。

3）Desk类

（1）定义一个Desk类（书桌类）。

（2）为Desk类定义一个String型的color成员变量，double型的length、width、height成员变量。

（3）为Desk类定义一个计算价格的方法calculate()，要求根据书桌的高度、宽度、长度计算书桌的售价，售价＝（高度×宽度×长度）/1 000，该方法返回书桌的售价。

（4）创建一张书桌（黑色，长度＝100，宽度＝60，高度＝80），调用calculate()方法输出售价。

[illegible]

[illegible] public void [illegible] B. public class Mocel ( ) [illegible]

C. public class void number [illegible] D. public class Car [illegible]

[illegible]

[illegible] double [illegible]

[illegible]

[illegible] double [illegible] double [illegible]

[illegible]

[illegible] static [illegible]

[illegible] public [illegible]

[illegible]

第4章

# 面向对象（二）

——有志者，事竟成

# 4.1 面向对象继承性

## 学习目标

（1）了解继承、包的概念。

（2）掌握类的继承、方法的重写。

（3）掌握 this 关键字、super 关键字、final 关键字的使用方法。

（4）掌握访问控制修饰符的使用方法。

## 相关知识

### 4.1.1 继承

#### 一、继承的概念

世界万物存在千丝万缕的联系，它们存在共同的特性，也有各自的特异性。图 4-1-1 所示的继承关系如下；人类（姓名、性别、年龄）；教师类（姓名、性别、年龄、职称）；学生类（姓名、性别、年龄、学号）。

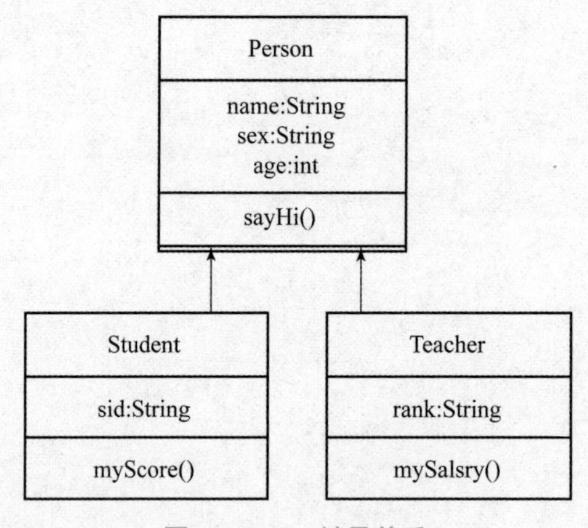


图 4-1-1 继承关系

教师类和学生类拥有人类的所有属性（姓名、性别、年龄），同时它们又拥有自己特有的属性（职称和学号）。因此，教师类继承人类，学生类继承人类。

在 Java 中，类的继承就是构建一个新类，新类除了有自己的属性和行为外，还从已有类获得属性和行为。新类称为子类或派生类，已有类称为父类或基类。

小贴士

Java 只支持单继承，但允许多层继承，如图 4-1-2 所示。

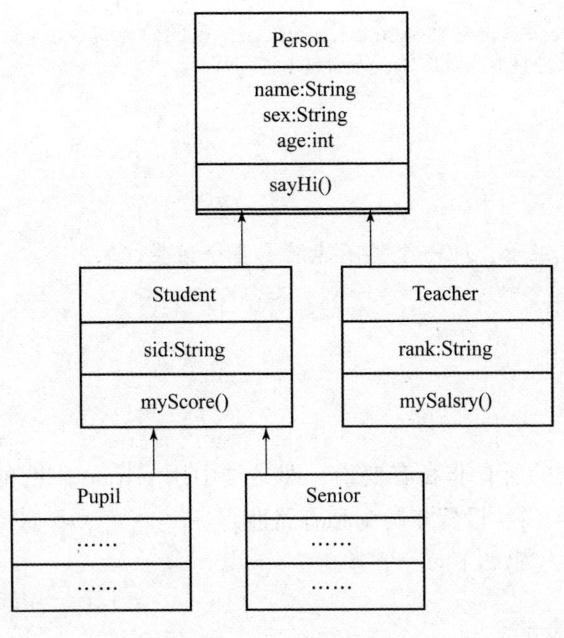


图4-1-2　多层继承关系

## 二、继承的定义

定义子类（派生类）的语法格式如下：

```
 class  子类名  extends  父类名{
//子类成员变量
//子类成员方法
}
```

【例4-1-1】定义父类Person，代码如下：

```
1  public class Person {
2     //成员变量
3     public String name , sex; //姓名、性别
4     public int age; //年龄
5     //成员方法
6     public void sayHi(){System. out .println("Hi");}
7  }
```

定义子类Student，代码如下：

```
1  public class Student extends Person {
2     //1. 子类自己的成员变量
3     String sid;
4     //2. 默认继承父类的成员变量:name、sex、age
5     //3. 子类自己的成员方法
```

```
6    public void myScore(){System. out .println("My score is very good!");}
7    //4. 默认继承父类的成员方法:sayHi()
8  }
```

Java 不支持多重继承，即一个类不能继承多个父类。

## 三、属性的继承与隐藏

### 1. 属性的继承

子类可以继承父类的所有非私有变量，即父类中用 private 声明的成员变量无法被子类继承。修改【例 4-1-1】，设置年龄为私有属性。

定义父类 Person，代码如下：

```
1  public class Person {
2      public String name;
3      public String sex;
4      private int age; //修改为私有属性
5      public void sayHi(){System. out .println("Hi");}
6  }
```

定义子类 Student，代码如下：

```
8  public class Student extends Person {
9      String sid;
10      public void myScore(){System. out .println("My score is very good!");}
11      public static void main(String[ ] args) {
12          Student s = new Student();
13          System. out .println(s.age);
14      }
15  }
```

运行结果如图 4-1-3 所示。

```
Exception in thread "main" java.lang.Error: Unresolved compilation problem:
        The field Person.age is not visible
        at chapter01.Student.main(Student.java:13)
```

图 4-1-3　运行结果

【说明】

程序出错，说明子类 Student 无法继承父类 Person 的私有属性 age。

### 2. 属性的隐藏

所谓属性的隐藏，是指子类拥有与父类同名的变量。当子类调用父类方法时，操作的是继承自父类的变量；而当子类调用它自己定义的方法时，所操作的就是它自己定义的变量，而把继承自父类的同名变量“隐藏”起来。

修改【例4-1-1】在子类Student中定义性别sex为char类型，这样父类和子类拥有一个同名变量sex，但两者的数据类型不同。代码如下：

```
1  public class Student extends Person {
2      String sid;
3      char sex; //与父类变量同名,但数据类型不同
4      public void myScore(){System.out.println("My score is very good!");}
5      public static void main(String[] args) {
6          Person p = new Person();
7          p.sex = "男";
8          Student s = new Student();
9          s.sex = '女';
10         System.out.println("父类 sex:" + p.sex);
11         System.out.println("子类 sex:" + s.sex);
12     }
13 }
```

运行结果如图4-1-4所示。

父类sex:男
子类sex:女

图4-1-4 运行结果

## 四、方法的继承与重写

### 1. 方法的继承

**父类中的非私有方法都可以被子类继承，即父类中用private声明的成员方法无法被子类继承。修改【例4-1-1】中的父类Person，增加私有方法showPerson()，代码如下：**

```
1  public class Person {
2      public String name;
3      public String sex;
4      private int age;
5      //私有成员方法
6      private void showPerson() {
7          System.out.println("我的名字:" + name);
8      }
9      public void sayHi(){System.out.println("Hi");}
10 }
```

子类Student继承父类中的所有非私有方法，代码如下：

```
4  public class Student extends Person {
5      String sid;
6      char sex; //与父类变量同名,但数据类型不同
7      public void myScore(){System.out.println("My score is very good!");}
8      public static void main(String[] args) {
9          Student s = new Student();
10          s.name = "王萌";
11          s.sayHi();
12          s.showPerson();
13     }
14 }
```

运行结果如图 4－1－5 所示。

```
Exception in thread "main" java.lang.Error: Unresolved compilation problem:
	The method showPerson() from the type Person is not visible

	at chapter01.Student.main(Student.java:12)
```

图 4－1－5　运行结果

【说明】

程序出错，说明子类 Student 无法继承父类 Person 的私有方法 showPerson()。

**2. 重写父类的方法**

在继承关系中，子类需要对继承的方法进行功能修改，即对父类的方法进行重写（复写）。

修改【例 4－1－1】，重写父类 Person 的 sayHi()方法，代码如下：

```
1  public class Student extends Person {
2      String sid;
3      char sex;
4      public void myScore(){System.out.println("My score is very good!");}
5      //重写,方法名、参数、返回值数据类型均与父类相同
6      public void sayHi(){
7          System.out.println("HI,my name is " +name);
8      }
9      //主函数
10      public static void main(String[] args) {
11         Student s = new Student();
12         s.name = "王萌";
13         s.sayHi();
14      }
15 }
```

运行结果如图 4－1－6 所示。

```
HI,my name is 王萌
```

图 4－1－6　运行结果

在子类中重写父类的成员方法时，应注意以下几点。

（1）子类的方法与父类的方法完全相同（返回值数据类型、方法名、参数列表）。

（2）子类的方法不能缩小父类的方法的访问权限。

（3）子类无法继承父类的私有方法，也就谈不上重写。

（4）静态方法不能被重写。

## 五、this 关键字、super 关键字、final 关键字

this 和 super 关键字是常用于指代子类对象和父类对象的关键字。

**1. this 关键字**

this 关键字用于引用自身对象，主要有以下两种使用方式。

（1）方法的形参与类的成员变量同名，例如：

```
1  public class Person {
2      public String name ,sex;
3      public int age;
4      public void setPerson(String name,String sex, int age) {
5          this.name = name;//参数与成员变量同名
6          this.sex = sex;
7          this.age = age;
8      }
9  }
```

（2）构造方法只能在创建对象时由系统自动调用，不允许像调用成员方法一样调用它，但允许在一个构造方法中调用另外一个重载的构造方法，此时不能通过构造方法名来调用，而是用 this（参数列表）的形式。例如，

```
1  public class Person {
2      public String name ,sex;
3      public int age;
4      //构造方法
5      public Person(String name) {
6          this.name = name;
7      }
8      public Person(String name, String sex) {
9          this(name);//调用上面带 name 成员变量的构造方法
10         this.sex = sex;
11      }
12  }
```

2. super 关键字

super 关键字指代当前对象的直接父类对象，主要有以下 3 种使用方式。

（1）super. 变量名：访问父类中被隐藏的成员变量。

（2）super. 方法名（［参数表］）：调用父类中被重写的方法。

（3）super（［参数表］）：调用父类的构造方法，此时可用 super 关键字表示父类的构造方法。

例如：

```
1  public class Student extends Person{
2      String sid;
3      char sex;
4      public void myScore(){System.out.println("My score is very good!");}
5      public void sayHi(){
6          System.out.println("HI,my name is " +name);
7      }
8      //构造方法
9      public Student(String name, String sex, String sid) {
10          super(name,sex);//调用父类的构造方法
11          this.sid = sid;
12          super.sayHi();//调用父类的 sayHi()方法
13          System.out.println(super.sex);//调用父类 String 型的成员变量 sex
```

```
14      }
15      //主函数
16      public static void main(String[] args) {
17          Student s = new Student("王萌","男","22001");
18          }
19      }
```

### 3. final 关键字

final 是“最终”的意思，也就是说，被 final 关键字修饰的类不允许被继承，方法不允许被重写，变量不允许被再次赋值，具体如下。

(1) 被 final 关键字修饰的类不允许被继承。

(2) 被 final 关键字修饰的方法不允许被重写。

(3) 被 final 关键字修饰的变量其实是一个常量，不允许被再次赋值。

## 4.1.2 包

Java 引入包（package）机制，以便 Java 类的组织与管理。包具有类似文件夹的功能，用于实现对文件的管理，需要注意的是包名应为小写字母。

### 一、声明包

使用 package 语句在源文件第 1 行声明包，例如：

```
package chapter04; //必须在源文件第 1 行
public class Person {
      .....
}
```

### 二、使用包

在程序开发中，不同包中的类经常需要互相调用，调用方法有以下两种。

(1) 每次调用时必须给出完整的类名，即“包名 . 类名”的形式，如图 4 - 1 - 7 所示。

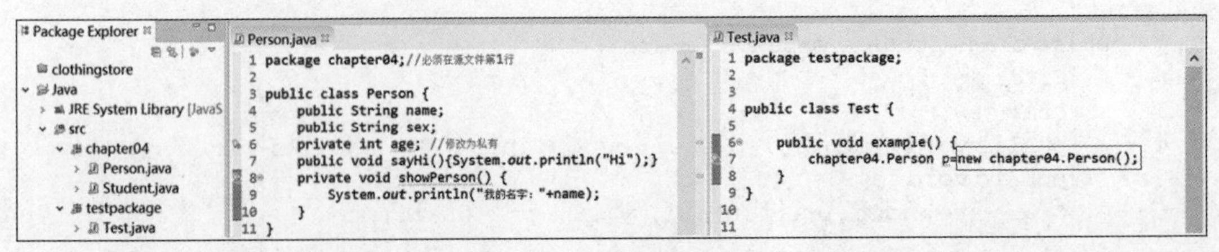


图 4 - 1 - 7 调用包（1）

(2) 通过 import 语句事先导入包中的类，每次调用时不必给出完整的类名。若需要用到一个包中的多个类，则可以采用“import 包名 . *”的方式一次性导入包中的所有类。需要注意的是 import 语句必须位于 package 语句之后、类定义之前，如图 4 - 1 - 8 所示。

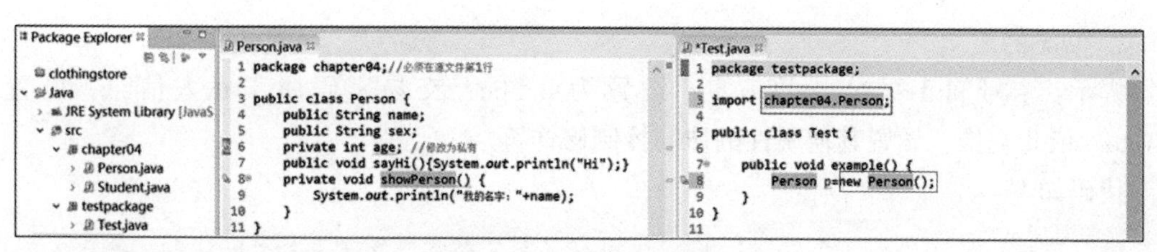


图 4-1-8　调用包（2）

## 4.1.3　访问控制修饰符

Java 针对类、成员变量和成员方法提供了 4 种访问控制级别，分别是 public、protected、default、private，它们的访问范围从大到小，如图 4-1-9 所示。

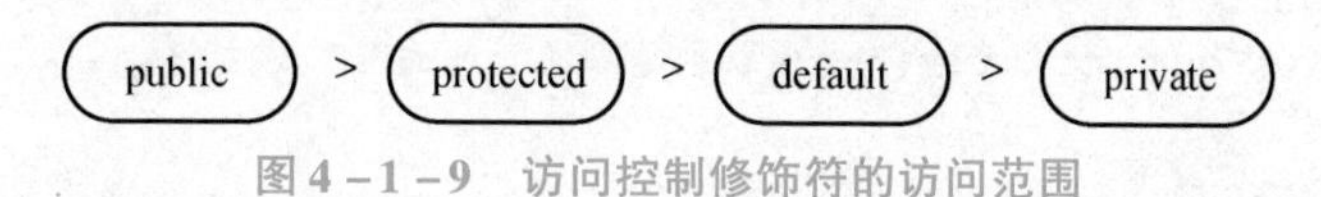


图 4-1-9　访问控制修饰符的访问范围

（1）public：公共访问级别，用 public 修饰的类或类的成员允许被所有类访问。

（2）protected：子类访问级别，用 protected 修饰的成员允许被本包中的其他类访问，也可以被不同包中该类的子类访问。

（3）default：包访问级别，当类或类的成员不使用任何访问控制修饰符时，则为默认的访问控制修饰符，这样的类或类的成员只允许被本包中的其他类访问。

（4）private：类访问级别，用 private 修饰的成员只能被类中的其他成员访问，其他类则无法直接访问。private 很好地实现了类的封装性。

表 4-1-1 直观地展示了 4 个访问控制修饰符的访问范围。

表 4-1-1　访问控制修饰的访问范围

| 访问范围 | public | protected | default | private |
| --- | --- | --- | --- | --- |
| 同类 | √ | √ | √ | √ |
| 同包 | √ | √ | √ | — |
| 子类 | √ | √ | — | — |
| 所有类 | √ | — | — | — |

**案例**

银行有两种银行卡，分别是普通银行卡和 VIP 银行卡。两类银行卡存在怎样的关系？在 Java 中如何体现它们的关系？如何处理银行卡中的隐私数据？

【分析】

（1）VIP 银行卡是特殊的银行卡，拥有普通银行卡的所有特性，因此二者存在继承关系。

（2）银行卡信息中关于密码、余额之类的数据应该设置为 protected，不能设置为 private，否则 VIP 银行卡无法继承这两个成员变量。

（3）chapter04. card 包：银行卡类。

### 1. 父类的定义

提示：卡号属于可公开的信息，设置为 public；交易密码属于私人信息，设置为 private。依此类推，设置其他属性的访问控制修饰符。

代码如下：

```
public class Card {
    public String cardID , customerName , password;
    protected double money;
    double queryMoney(){
        return money;
    }
    Protected void drawMoney(double m) { //取钱
      if (m >500000)
          System. out .println("一天取款总额不可超过50 万元");
    else {
        money =money - m;
        System. out .println("您的余额为:" +money);
    }
  }
}
```

### 2. 子类的定义

代码如下：

```
class VIPCard extends Card{
    public double loanLimit , lendingRate;//贷款额度、贷款利率
    protected void drawMoney(double m) { //取款
        money =money - m;
        System. out .println("您的余额为:" +money);
    }
  }
```

### 3. 测试类设计

代码如下：

```
public class Test{
    public static void main(String[] args) {
        VIPCard vip =new VIPCard();
        vip.cardID ="387987987987970979 7";
        vip.money =100000;
        vip.loanLimit =1000000;
        vip.drawMoney(10000);
        Card card =new Card();
        card.money =1000000;
        card.drawMoney(600000);
    }
  }
```

## 任务工单

<table>
<tr><td>任务名称</td><td colspan="5">图书信息管理系统——旧图书类（Old Book）</td></tr>
<tr><td>班级</td><td></td><td>学号</td><td></td><td>姓名</td><td></td></tr>
<tr><td>任务要求</td><td colspan="5">（1）定义旧图书类 OldBook，该类继承 Book 类。<br>（2）OldBook 类成员变量：旧书描述（describe;）。<br>（3）VIPCustomer 类复写 queryBook( ) 方法，根据图书编号，输出该编号图书的书名、作者及定价信息。<br>（4）根据 4 个访问控制修饰符的访问范围，合理设置 Book 类和 OldBook 类各成员变量及方法的访问控制范围。<br>（5）测试完善后的 Book 类和 OldBook 类。</td></tr>
<tr><td>任务实现</td><td colspan="5"></td></tr>
<tr><td>异常及待解决问题记录</td><td colspan="5"></td></tr>
</table>

# 4.2 面向对象多态性

## 学习目标

(1) 掌握抽象类、抽象方法的使用方法。
(2) 掌握接口的定义。
(3) 掌握实现接口的方法。
(4) 掌握多态的使用方法。

## 相关知识

### 4.2.1 抽象类

在类的定义中，使用成员方法描述类的行为，但在有些情况下，这些行为无法具体实现。例如在形状类 Shape 中，由于不知道具体形状，所以无法确定面积计算方法 area()。在 Java 中，使用抽象方法表示这种当前无法实现的方法。

**【例 4-2-1】**代码如下：

```
1  public abstract class Shape {//抽象类
2    public   abstract void area();//抽象方法
3  }
```

#### 一、抽象方法

抽象方法的语法格式如下：

```
abstract 返回值数据类型  抽象方法名(参数列表)；
```

(1) 必须用 abstract 关键字修饰。
(2) 仅有方法头而没有方法体，即不能出现大括号。
(3) 必须以“;”结束。
(4) 不能使用 private 访问控制修饰符，因为抽象方法必须被子类实现，若使用 private 则子类无法继承该方法。

#### 二、抽象类

抽象类的语法格式如下：

```
abstract class   类名{
  ……
}
```

（1）抽象类必须用 abstract 关键字修饰。

（2）抽象类不一定包含抽象方法，但是，包含抽象方法的类一定是抽象类。

（3）抽象类可以包含非抽象方法。

（4）抽象类不能被实例化，主要用来被继承，因为抽象类可以包含抽象方法，抽象方法没有方法体，无法被调用。

## 三、抽象类的使用

要调用抽象类的方法，则需要创建一个子类继承抽象类，并在子类中实现抽象类的所有抽象方法，通过子类对象访问抽象类方法。

**【例4-2-2】**代码如下：

```
1  //抽象类
2  public abstract class Shape{
3      public String name;
4      //成员方法
5      public void print(){System. out .println("my name is " +name);}
6      //抽象方法(求周长、求面积)
7      public abstract double area();
8      public abstract double perimeter();
9  }
10 //子类继承抽象类,实现抽象类中的所有方法
11 class Circle extends Shape{
12     double radius;
13     @ Override//重写求周长、求面积方法
14     public double area() {
15         return 3.14 * radius * radius;
16     }
17     @ Override
18     public double perimeter() {
19         return 2 * 3.14 * radius;
20     }
21 }
22 //测试类
23 public class Test {
24     public static void main(String[] args) {
25         Circle c1 =new Circle();
26         c1.radius =1;
27         System. out .println("圆面积:" +c1.area() +",周长:" +c1.perimeter());
28     }
29 }
```

运行结果如图4-2-1所示。

```
圆面积：3.14,周长：6.28
```

图4-2-1　运行结果

### 4.2.2 接口

如果一个类中的所有方法都是抽象的，那么可以用另外一种方式定义这个类，即接口。

#### 一、接口的定义

定义接口的语法格式如下：

```
  [public] interface 接口名 [extends 父接口列表]{
  [public][static][final]类型 变量名=初始值;//常量域声明
  [public][abstract]返回值类型 方法名(参数列表);//抽象方法声明
}
```

（1）使用 interface 关键字定义接口。

（2）接口中的所有方法都是抽象的，即不允许出现非抽象方法。

（3）接口允许多继承，即“interface 接口1 extends 接口2，接口3…。”

（4）如果在子接口中定义了和父接口同名的变量或方法，则子接口的变量或方法覆盖父接口中的变量或方法。

（5）接口中的变量默认使用“[public][static][final]”来修饰，即全局常量。

（6）接口中的方法默认使用“[public][abstract]”来修饰，即抽象方法。

（7）因为接口中有抽象方法，所以接口不能被实例化，只能被实现。

【例4-2-3】代码如下：

```
1  //父接口
2  public interface Interf01 {
3      public abstract double area();
4  }
5  public interface Interf02 {
6      public abstract double perimeter();
7  }
8  //子接口继承两个父接口
9  interface Interf03 extends Interf01, Interf02{
10     public abstract void shapeName();
11 }
```

【说明】

父接口 Interf 01 包含求面积的抽象方法 area()，父接口 Interf02 包含求周长的抽象方法 perimeter()；子接口 Interf03 自带抽象方法 shapeName()，及继承自父接口的抽象方法 area()、perimeter()。

#### 二、接口的实现

由于接口中的方法都是抽象方法，无法通过实例化对象的方式调用这些方法，所以必须定义一个实现类，并使用 implements 关键字实现接口中的所有方法，这样就可以通过实现类

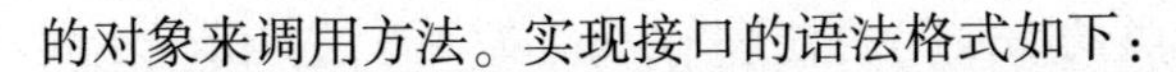
的对象来调用方法。实现接口的语法格式如下：

```
class  实现类  implements  接口1,接口2…{
  //实现接口1,接口2…中的所有抽象方法
}
```

（1）接口不能被实例化，因为它包含抽象方法。

（2）实现类可以是抽象类，也可以是非抽象类。如果实现类是抽象类，则可以只实现接口中的部分抽象方法；如果实现类是非抽象类，则必须实现接口中的所有抽象方法。

（3）实现类通过implements 可以实现多个接口，接口之间用逗号隔开。

（4）实现类中所有实现了的抽象方法都必须显式地使用 public 访问控制修饰符，否则将被警告缩小接口中方法的访问范围。

（5）实现类同时继承一个类和实现一个接口时，extends 关键字必须在implements 关键字之前。

**【例4－2－4】** 代码如下：

```
1  public class Circle implements Interf01,Interf02{
2      double radius;
3      @ Override
4      public double area() {
5          return 3.14 * radius * radius;
6      }
7      @ Override
8      public double perimeter() {
9          return 2 * 3.14 * radius;
10     }
11   }
12   //测试类
13   public class Test {
14       public static void main(String[] args) {
15       Circle c1 = new Circle();
16       c1.radius = 1;
17       System. out .println("圆面积:" + c1.area() + ",周长:" + c1.perimeter());
18     }
19   }
```

运行结果如图4－2－2 所示。

圆面积：**3.14,**周长：**6.28**

图4－2－2　运行结果

## 4.2.3　多态

### 一、多态概述

多态性是面向对象程序设计的重要特性之一，它与封装性和继承性一起构成了面向对象程序设计的三大特性。在面向对象程序设计中，多态性主要体现在向一个方法中传递不同的

参数，可以得到不同的响应消息。例如形状类的 area( )方法和 perimeter( )方法，由于不同形状的计算方法不同，所以可以为 area( )方法和 perimeter( )方法传一个形状型的参数。若传入参数为圆类对象，则按照求圆面积和圆周长的方法进行计算；若传入参数为正方形对象，则按照求正方形面积和正方形周长的方法进行计算。

**【例 4-2-5】** 代码如下：

```
1  interface Shape {
2      public abstract double area();
3      public abstract double perimeter();
4  }
5  class Circle implements Shape{
6      double radius;
7      //重写求周长、求面积方法
8      @ Override
9      public double area() {
10         return 3.14 * radius * radius;
11     }
12     @ Override
13     public double perimeter() {
14         return 2 * 3.14 * radius;
15     }
16 }
17 class Square implements Shape{
18     double length;
19     //重写求周长、求面积方法
20     @ Override
21     public double area() {
22         return length * length;
23     }
24     @ Override
25     public double perimeter() {
26         return 4 * length;
27     }
28 }
29 public class Test {
30     public   void printResult(Shape s) {
31         System. out .println("面积:" +s.area() +"周长:" +s.perimeter());
32     }
33     public static void main(String[] args) {
34         Shape c1 =new Circle();
35         Shape s1 =new Square();
36         Test t =new Test();
37         t. printResult (c1);
38         t. printResult (s1);
39     }
40 }
```

运行结果如图 4-2-3 所示。

```
面积：0.0,周长：0.0
面积：0.0,周长：0.0
```

图 4-2-3　运行结果

【说明】

从运行结果分析，两次面积和周长均为 0.0，返回值的数据类型是正确的，因此问题出现在程序中没有为圆的半径赋值，也没有为正

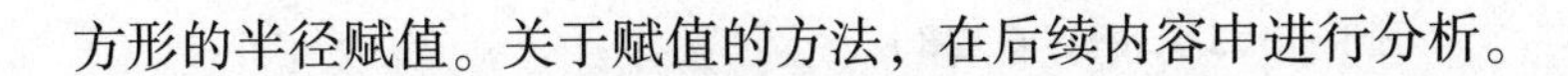

方形的半径赋值。关于赋值的方法，在后续内容中进行分析。

## 二、子类对象自动转换成父类类型

子类对象自动转换成父类类型的语法格式如下：

```
父类 父类对象名 =new  子类();
```

（1）子类对象转换成父类类型是自动的，无须显式声明。

（2）子类对象转换成父类类型后，不允许访问子类的变量及方法。

【例4-2-5】中子类对象c1要转换成父类Shape类型，采用“Shape c1 = new Circle( );”的方式，子类对象c1自动转换为Shape类型。为了解决上述半径为0的问题，可以采用“c1. radius =1”的方式来完成赋值吗？答案是不行，因为c1已经转换为父类对象，所以不允许再访问子类变量及方法。

## 三、父类对象强制转换成子类类型

父类对象强制转换成子类类型的语法格式如下：

```
子类 子类对象名 =(子类)父类对象名;
```

对于【例4-2-5】中父类对象无法访问子类变量和方法问题，采用强制转换的方法，把父类对象强制转换成子类类型，通过子类对象访问其变量和方法。【例4-2-5】修改如下：

```
1  public class Test {
2      public  void printResult(Shape s) {
3          //instanceof 关键字可以判断一个对象是否为某个类(或接口)的实例或子类实例
4          if (s instanceof Circle) {
5              Circle c =(Circle)s;
6              c.radius =1;
7              System. out .println("面积:" +c.area() +",周长:" +c.perimeter());
8          } else {
9              Square c =(Square)s;
10             c.length =1;
11             System. out .println("面积:" +c.area() +",周长:" +c.perimeter());
12         }
13     }
14     public static void main(String[] args) {
15         Shape c1 =new Circle();
16         Shape s1 =new Square();
17         Test02 t =new Test02();
18         t.printResult(c1);
19         t.printResult(s1);
20     }
21 }
```

运行结果如图4-2-4所示。

```
面积：3.14,周长：6.28
面积：1.0,周长：4.0
```

图 4-2-4　运行结果

**案例**

设计员工工资管理模块，工资主要按员工级别计算，高职称级别员工基本工资为 8 000 元，加班日工资为 150 元，请假日扣除工资 100 元；低职称级别员工基本工资为 5 000 元，加班日工资为 100 元，请假日扣除工资 50 元。基本工资及额外工资的计算因人而异。

【分析】

(1) 由于员工职称无法确定，无法确定基本工资、额外工资、总工资的计算方法，所以定义为抽象方法。

(2) 封装抽象方法的可以是抽象类，也可以是接口，这里选择接口实现。

**1. 接口的设计**

1) 父接口

(1) BaseWage 接口：计算基本工资的接口。

定义计算基本工资的抽象方法。

(2) ExtraWage 接口：计算额外工资的接口。

①定义计算请假扣除工资的抽象方法。

②定义计算加班补贴工资的抽象方法。

2) 子接口

Wage 接口：继承上面两个父接口。

定义计算最终工资的抽象方法。

**2. 接口的定义**

代码如下：

```
interface BaseWage{
    double wage();
}
interface ExtraWage{
    double askForLeaveWage();
    double workOverWage();
}
public interface Wage extends BaseWage,ExtraWage{
    double totalWages();
}
```

**3. 接口的实现**

设计员工类 WorkMan 作为实现类，实现员工工资的计算。代码如下：

```
public class WorkMan extends Person implements Wage{
    private int month;
    private int askForLeaveDays,workOverTimeDays;
    private String rank;
    @ Override
```

```
    public double wage() {
        if (rank.equals("high"))
            return 8000;
      else
            return 5000;
    }
    @ Override
    public double askForLeaveWage() {
        if (rank.equals("high"))
            return askForLeaveDays * 100;
        else
            return askForLeaveDays * 50;
    }
    @ Override
    public double workOverWage() {
        if (rank.equals("high"))
            return askForLeaveDays * 80;
          else
            return askForLeaveDays * 40;
    }
    @ Override
    public double totalWages() {
            return wage() + workOverWage() - askForLeaveWage();
    }
    public WorkMan(int month, int askForLeaveDays, int workOverTimeDays, String rank ) {
            super ();
            this .askForLeaveDays = askForLeaveDays;
            this .workOverTimeDays = workOverTimeDays;
            this .rank = rank;
            this .month = month;
        }
    public WorkMan() {
            super ();
        }
}
```

4. 测试类的设计

代码如下：

```
public class Test{
    public static void main(String[] args) {
        WorkMan w = new WorkMan(1,1,2,"high");
        System. out .println("我" + w.month + "月份的工资是:"
                + w.totalWages() + ";扣了:" + w.askForLeaveWage() + ";加班费:"
                + w.workOverWage());
    }
}
```

## 任务工单

<table>
<tr><td>任务名称</td><td colspan="5">图书信息管理系统——图书、杂志业务处理</td></tr>
<tr><td>班级</td><td></td><td>学号</td><td></td><td>姓名</td><td></td></tr>
<tr><td>任务要求</td><td colspan="5">（1）定义杂志类 Magazine，该类包含杂志名（magazineName）、期次（period）。<br>（2）定义抽象类 Setting，该类包含 1 个抽象方法 int MaxBorrowDays( )，用于设置最长的借阅时间。<br>（3）定义接口 Operator，该接口包含 2 个抽象方法 boolean renewBook( ) 及 boolean preBook( )，分别用于设置是否允许续借及预约。<br>（4）Book 类、Magazine 类分别继承抽象类 Setting 及接口 Operator。实现如下设置：Book 类最长借阅时间为 30 天，允许续借，允许预约；Magazine 类最长借阅时间为 15 天，不允许续借，不允许预约。</td></tr>
<tr><td>任务实现</td><td colspan="5">[完整记录两个类（Book、Magazine）、抽象类、接口的定义及实现]</td></tr>
<tr><td>异常及待解决问题记录</td><td colspan="5"></td></tr>
</table>

## 练习题

### 1. 选择题

（1）在类的继承关系中，需要遵循（　　）继承原则。

A. 多重　　B. 单一　　C. 双重　　D. 不能继承

（2）关于 super 关键字，以下说法正确的是（　　）。（多选）

A. super 关键字可以调用父类的构造方法

B. super 关键字可以调用父类的普通方法

C. super 关键字与 this 关键字不能同时存在于一个构造方法中

D. super 关键字与 this 关键字可以同时存在于一个构造方法中

（3）以下说法正确的是（　　）。（多选）

A. Java 允许一个类实现多个接口

B. Java 不允许一个类继承多个类

C. Java 允许一个类同时继承一个类并实现一个接口

D. Java 允许一个接口继承一个接口

（4）当类中的一个成员方法被（　　）访问控制修饰符修饰时，该方法只能在本类中被访问。

A. public　　B. protected　　C. private　　D. default

（5）关于抽象类的说法正确的是（　　）。（多选）

A. 抽象类中可以有非抽象方法

B. 如果父类是抽象类，则子类必须重写父类的所有抽象方法

C. 不能用抽象类创建对象

D. 接口和抽象类是同一个概念

### 2. 填空题

（1）在程序开发中，要将一个包中的类导入当前程序，可以使用______关键字。

（2）一个类可以从其他类派生，派生的类称为______，用于派生的类称为______或者__________。

（3）在 Java 中，所有类都直接或间接继承__________类。

（4）一个类如果实现一个接口，那么它就需要实现接口中定义的全部________，否则该类就必须被定义成________。

（5）在 Java 中定义类时，如果前面使用________关键字修饰，则该类不可以被继承。

### 3. 设计题

1）继承

（1）父类定义。

①定义一个 Circle 类表示圆。

②为 Circle 类定义 private double 型的成员变量 radius（半径）。

③为 radius 定义访问方法和设置方法。

④为 Circle 类定义一个不带参数的构造方法和一个带 radius 的构造方法。

⑤定义成员方法 double perimeter( )，该方法不带参数，作用是返回圆的周长。

⑥定义成员方法 double area( )，该方法不带参数，作用是返回圆的面积。

（2）子类定义。

①定义一个 Cylinder 类表示圆柱，使 Cylinder 类继承 Circle 类。

②为 Cylinder 类定义 double 型的成员变量 height（高）。

③为 Cylinder 类定义构造方法。

④重写 double area( )方法，使该方法用于求圆柱的表面积。

2）抽象类及接口

（1）创建一个打折的抽象类 DaZhe，该类包含一个 double 型的 discount( )（打折）抽象方法。

（2）创建一个售后服务的接口 ShouHou，该接口包含一个 void 型的 back( )（退货）抽象方法和一个 void 型的 change( )（换货）抽象方法。

（3）新建一个 Fridge 类代表冰箱，该类包含一个 String 型的 version（型号）、double 型的 price（单价），该类继承抽象类 dazhe，并实现打折功能：如果单价超过 5 000 元，则打 5 折；如果单价低于 5 000 元，则打 7 折。

（4）Fridge 类实现 ShouHou 接口，要求：如果为退货，则输出“我要退货”；如果为换货，则输出“我要换货”。

# 第5章

# 图形用户界面（GUI）

——骐骥一跃，不能十步

# 5.1 容器与组件

## 学习目标

（1）了解 Swing 的容器与组件。

（2）掌握常用容器的使用方法。

（3）掌握常用布局方式。

（4）掌握按钮、标签、文本框、单选按钮、复选框、列表框等图形组件的创建和操作方法。

## 相关知识

### 5.1.1 AWT 与 Swing

#### 一、AWT 与 Swing

抽象窗体工具包（Abstract Windowing Toolkit，AWT）是自 JDK1.0 就推出的构建 GUI 的标准 API，习惯上称为重量级组件。其工作原理是把 GUI 组件创建及交互工作交给目标平台的本地 GUI 工具箱完成。这就存在以下缺点：GUI 的外观在不同平台上可能不同，以及需要消耗大量的系统资源来解决平台组件不兼容的问题等。

在 JDK1.2 中，AWT 组件在很大程度上被 Swing 组件取代，Swing 组件称为轻量级组件。其工作原理是把 GUI 组件创建及交互工作交给由纯 Java 编写的 UI 类完成，这些 UI 类被加载到 Java 的运行环境中，因此组件的外观不依赖平台，不受兼容性的影响，可提高系统性能。Swing 组件是 AWT 组件的一部分，组件名称是在 AWT 类库中相同功能的组件名称前加字母 J。

#### 二、常用 Swing 组件

如图 5-1-1 所示，Swing 组件包括容器及组件。容器本身也是组件，其作用是容纳其他组件（可以是容器，也可以是组件），就像一口箱子可以容纳另一口箱子，也可以容纳其他物品。Java 中定义的 GUI 顶级窗口一定是一个容器（例如 JFrame 窗体），其他组件（例如 JButton 按钮）则放置在容器中。

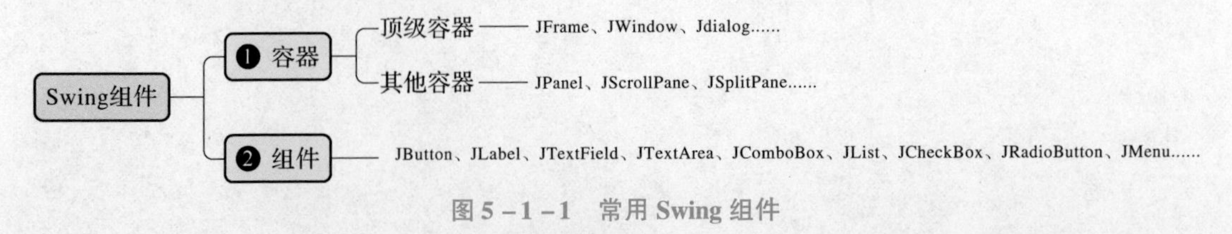


图 5-1-1 常用 Swing 组件

## 5.1.2　常用容器

容器的主要作用是组织其他组件，并按照一定的方式排列，组件随容器的打开而打开。容器常用 add()方法添加其他组件，setLayout()方法设置容器中组件的摆放方式。常用容器有 JFrame、JPanel、JScrollPane、JsplitPane。

### 一、JFrame

JFrame 类继承 java. awt. Frame 类，是一个顶级容器，其默认的布局方式是边界布局（BorderLayout）。

**1. JFrame 类的常用构造函数**

（1）JFrame（）：创建一个不带标题的窗体。

（2）JFrame（String title）：创建一个带标题的窗体。

**2. JFrame 的常用方法**

JFrame 的常用方法如表 5 – 1 – 1 所示。

表 5 – 1 – 1　JFrame 的常用方法

| 方法 | 功能说明 |
| --- | --- |
| Component add() | 将组件添加到窗体中 |
| void setLayout() | 设置窗体的布局方式 |
| void setSize(int width, int height) | 设置窗体的大小（宽和高） |
| void setLocation(int x, int y) | 设置窗体左上角的坐标 |
| void setBounds(int x,int y,int w,int h) | 设置窗体左上角的坐标，修改窗体大小 |
| void setVisible(boolean b) | 设置窗体是否可见 |
| void setMenuBar(MenuBar m) | 设置窗体的菜单栏 |
| void setDefaultCloseOperation(JFrame. EXIT_ON_CLOSE) | 设置在关闭窗体的同时关闭后台应用程序 |

【例 5 – 1 – 1】代码如下：

```
1  public class P5_1_3 {
2      public static void main(String[] args) {
3          //创建一个窗体
4          JFrame frame = new JFrame("这是我的第一个窗体");
5          //组件(标签) 用来显示文字
6          JLabel label = new JLabel("hello",SwingConstants. CENTER );
7          frame.add(label); //把组件添加到窗体中
8          frame.setSize(800, 600); //设置窗体大小
9          frame.setVisible(true ); //窗体可视化
10         frame.setDefaultCloseOperation(JFrame. EXIT_ON_CLOSE); //关闭窗体进程
11     }
12 }
```

运行结果如图 5 – 1 – 2 所示。

## 二、JPanel

图 5-1-2　运行结果

JPanel 是一个无边框容器，它不是顶级容器，不能独立出现，必须放置在容器中，其默认的布局方式是流式布局（FlowLayout）。

JPanel 类的常用构造函数如下。

（1）JPanel()：创建一个具有流式布局的面板。

（2）JPanel（LayoutManager layout）：创建一个具有指定布局方式的面板。

**【例 5-1-2】**代码如下：

```
1  public class P5_1_4 {
2      public static void main(String[] args) {
3          JFrame frame  = new JFrame(); //创建窗体
4          JPanel panel = new JPanel(); //创建面板
5          panel.setBackground(Color. green ); //改变面板背景颜色
6          frame.add(panel); //把面板添加到窗体中
7          frame.setSize(600, 400);
8          frame.setVisible(true );
9          frame.setDefaultCloseOperation(JFrame. EXIT_ON_CLOSE );
10     }
11 }
```

### 5.1.3　常用布局方式

容器的布局方式主要用于控制组件在容器中的摆放位置，Java 中的常用布局有 BorderLayout、Flowlayout、GridLayout、GridBagLayout 等。每个容器都有其默认的布局方式，例如 JFrame 默认的布局方式为边界布局 BorderLayout，JPanel 默认的布局方式为流式布局 Flowlayout，可以通过 void setLayout(LayoutManager manager) 方法改变容器的布局方式。

## 一、流式布局

流式布局从上到下，从左到右依次放置组件，每行均居中，它是 JPanel、JApplet 默认的布局方式。Flowlayout 的常用构造方法如下。

（1）FlowLayout()：按默认居中方式放置组件。

（2）FlowLayout(int alignment)：按指定对齐方式放置组件。

（3）FlowLayout(int alignment,int h,int v)：按指定对齐方式放置组件。

参数 alignment 可以设置为 FlowLayout. RIGHT、FlowLayout. LEFT、FlowLayout. CENTER（默认值）；参数 h 设置每个组件左右间隔距离，参数 v 设置每个组件上下间隔距离，单位均为像素。

**【例 5-1-3】**代码如下：

```
1  public class FL extends JFrame{ //FL 默认的布局方式:边界布局
2          public FL () {
```

```
3           FlowLayout flowLayout = new FlowLayout(); //新建一个流式布局
4           this.setLayout(flowLayout); //窗体 FL 设置为流式布局
5           for(int i = 1;i <= 9;i ++) {
6               JButton b1 = new JButton("按钮" + i);
7               this.add(b1);
8           }
9           this.setSize(400, 300);
10          this.setVisible(true);
11          this.setDefaultCloseOperation(JFrame.EXIT_ON_CLOSE);
12      }
13      public static void main(String[] args) {
14          FL e = new FL();
15      }
16  }
```

运行结果如图5-1-3所示。

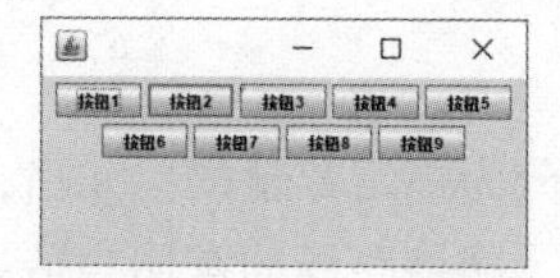

图5-1-3　运行结果

## 二、边界布局

边界布局将容器划分为东、南、西、北、中五个部分，通过指定方位来确定组件的放置位置，它是JFrame、JDialog默认的布局方式。边界布局的常用构造方法如下。

（1）BorderLayout()：按默认方式放置组件。

（2）BorderLayout(int h,int v)：指定组件间隔，参数h表示组件间左右间隔距离，参数v表示组件间上下间隔距离，单位均为像素。

**【例5-1-4】**代码如下：

```
1   public class BL extends JFrame{ //BL 默认的布局方式:边界布局
2       private JButton b1,b2,b3,b4,b5;
3       public BL () {
4           BorderLayout borderLayout = new BorderLayout(10,10); //新建边界布局
5           this.setLayout(borderLayout);
6           //添加组件
7           b1 = new JButton("北");
8           b2 = new JButton("南");
9           b3 = new JButton("西");
10          b4 = new JButton("东");
11          b5 = new JButton("中");
12          this.add(b1,BorderLayout.NORTH);
13          this.add(b2,BorderLayout.SOUTH);
14          this.add(b3,BorderLayout.WEST);
15          this.add(b4,BorderLayout.EAST);
16          this.add(b5,BorderLayout.CENTER);
17          this.setSize(400, 300);
18          this.setVisible(true);
19          this.setDefaultCloseOperation(JFrame.EXIT_ON_CLOSE);
```

```
20        }
21        public static void main(String[] args) {
22            BL e = new BL();
23        }
24    }
```

运行结果如图 5-1-4 所示。

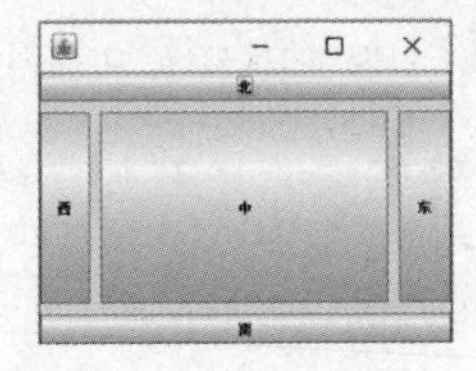

图 5-1-4　运行结果

## 三、网格布局（GridLayout）

网格布局按照表格样式将容器分为指定数量的单元格，每个单元格放置一个组件，组件放置位置依次从左向右，由上到下。网格布局的常用构造方法如下。

（1）GridLayout(int rows,int columns)：按指定行数和列数划分单元格。

（2）GridLayout(int rows,int columns,int h,int v)：按指定行数和列数划分单元格，并设置单元格间距。

**【例 5-1-5】**代码如下：

```
1   public class GL extends JFrame{ //GL 默认的布局方式:边界布局
2       public GL() {
3           GridLayout gridLayout = new GridLayout(3,3); //新建网格布局
4           this.setLayout(gridLayout); //设置窗体布局为网格布局
5           for(int i = 1;i <= 9;i ++) { //添加组件
6               JButton b = new JButton("b" + i);
7               this.add(b);
8           }
9           this.setSize(400, 300);
10          this.setVisible(true);
11          this.setDefaultCloseOperation(JFrame.EXIT_ON_CLOSE);
12      }
13      public static void main(String[] args) {
14          GL e = new GL();
15      }
16  }
```

运行结果如图 5-1-5 所示。

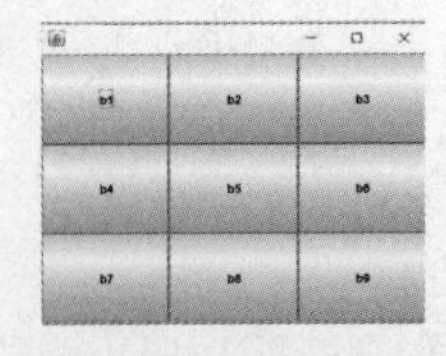

图 5-1-5　运行结果

### 5.1.4 常用组件

组件即在窗体中显示，能够与用户进行交互的对象。常用组件有 JLabel、JButton、JTextField、JTextArea、JComboBox、JRadioButton。

## 一、JLabel（标签）

JLabel 用来展示文本或者图片，用户只能看到文本，但不能修改文本。

**1. JLabel 类的常用构造方法**

（1）JLabel(String str)：创建指定文本的标签 。

（2）JLabel(String str,int align)：创建指定文本，指定对齐方式的标签。

（3）JLabel(Icon image)：创建指定图标的标签。

**2. JLabel 类的常用方法**

JLabel 类的常用方法如表 5 – 1 – 2 所示。

表 5 – 1 – 2　JLabel 类的常用方法

| 方法 | 功能说明 |
|---|---|
| String getText() | 获取标签显示的文本信息 |
| void setText(String s) | 修改标签显示的文本信息 |
| void setHorizontalAlignment(int alignment) | 设置标签文本水平方向对齐方式 |
| void setVerticalAlignment(int alignment) | 设置标签文本垂直方向对齐方式 |
| void setBackground(Color c) | 设置标签的背景颜色 |
| void setForeground(Color c) | 设置标签的字体颜色 |

## 二、JButton（按钮）

**1. JButton 类的常用构造方法**

（1）JButton()：创建一个无文本、无图标的按钮。

（2）JButton(String str)：创建一个带有文本的按钮。

（3）JButton(Icon icon)：创建一个带有图标的按钮。

（4）JButton(String str, Icon icon)：创建一个带有文本及图标的按钮。

**2. JButton 类的常用方法**

JButton 类的常用方法如表 5 – 1 – 3 所示。

表 5 – 1 – 3　JButton 类的常用方法

| 方法 | 功能说明 |
|---|---|
| String getText() | 获取按钮显示的文本信息 |

续表

| 方法 | 功能说明 |
| --- | --- |
| void setText( String s) | 修改按钮显示的文本信息 |
| void setHorizontalAlignment( int alignment) | 设置按钮文本水平方向对齐方式 |
| void setVerticalAlignment( int alignment) | 设置按钮文本垂直方向对齐方式 |
| void setICon( Icon icon) | 设置按钮显示的图标 |

## 三、JTextField（文本框）

JTextField 用于显示及编辑单行文本。

### 1. JTextField 类的常用构造方法

（1）JTextField( )：创建一个单行文本框。

（2）JTextField( String str)：创建一个带有初始文本的单行文本框。

（3）JTextField( int n)：创建一个指定列数的单行文本框。

### 2. JTextField 类的常用方法

JTextField 类的常用方法，如表 5 - 1 - 4 所示。

表 5 - 1 - 4　JTextField 类的常用方法

| 方法 | 功能说明 |
| --- | --- |
| String getText( ) | 获取单行文本框信息 |
| void setText( String s) | 修改单行文本框信息 |
| String getSelectedText( ) | 获取单行文本框中被选择的内容 |
| void setColumns( int n) | 设置单行文本框列数 |
| void setFont( ) | 设置单行文本框字体 |
| void setEditable( boolean b) | 设置单行文本框是否可以被编辑 |

【例 5 - 1 - 6】代码如下：

```
1  public class P5_1_10 extends JFrame {
2      private JPanel panel;
3      private JTextField textField,textField_1,textField_2;
4      public P5_1_10 () {
5          JLabel label = new JLabel("加法运算:");
6          textField = new JTextField();
7          JLabel label_1 = new JLabel(" + ");
8          textField_1 = new JTextField();
9          JButton button = new JButton(" = ");
```

```
10         textField_2 = new JTextField();
11         panel1 = new JPanel();
12         Panel2 = new JPanel();
13         panel1.add(label);
14         panel2.add(textField);
15         panel2.add(label_1);
16         panel2.add(textField_1);
17         panel2.add(button);
18         panel2.add(textField_2);
19         this. setLayout(new GridLayout(2,1));
20         this.add(panel1);
21         this.add(panel2);
22         //设置顶级容器的参数
23         this.setTitle("文本框、标签、按钮");
24         this.setDefaultCloseOperation(JFrame. EXIT_ON_CLOSE);
25         this.setVisible(true);
26     }
27     public static void main(String[] args) {
28         P5_1_10 p = new P5_1_10 ();
29     }
30 }
```

运行结果如图 5-1-6 所示。

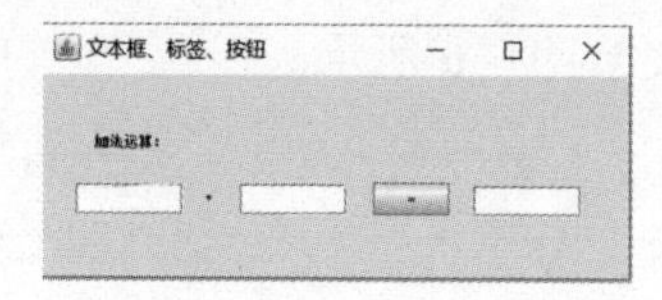


图 5-1-6 运行结果

## 四、JTextArea（文本区）

JTextArea 用于显示及编辑多行文本。

### 1. JTextArea 类的常用构造方法

（1）JTextArea()：创建一个多行文本框。

（2）JTextArea(int row,int columns)：创建一个指定行数及列数的多行文本框。

（3）JTextArea(String s,int row,int columns)：创建一个带有初始化文本并指定行数及列数的多行文本框。

### 2. JTextArea 类的常用方法

JTextArea 类的常用方法如表 5-1-5 所示。

表 5-1-5 JTextArea 类的常用方法

| 方法 | 功能说明 |
| --- | --- |
| String getText() | 获取多行文本框显示的信息 |

续表

| 方法 | 功能说明 |
| --- | --- |
| void setText(String s) | 修改多行文本框显示的信息 |
| String getSelectedText() | 获取多行文本框中被选择的内容 |
| void append(String s) | 将 s 追加到文本区中 |
| void setFont() | 设置多行文本框的字体 |

## 五、JCheckBox（复选框）

JCheckBox 用于实现多选功能，由多个 JCheckBox 组成一组选项，允许用户从中选择一个或者多个选项。

### 1. JCheckBox 类的常用构造方法

（1）JCheckBox()：创建一个无文本的复选框。

（2）JCheckBox(String s)：创建一个指定显示文本的复选框。

（3）JCheckBox(String s，boolean selected)：创建一个指定显示文本，并指定初始选定状态的复选框。

### 2. JCheckBox 类的常用方法

JCheckBox 类的常用方法如表 5-1-6 所示。

表 5-1-6　JCheckBox 类的常用方法

| 方法 | 功能说明 |
| --- | --- |
| boolean isSelected() | 判断复选框是否被勾选 |
| void setSelected(Boolean b) | 设置复选框是否被勾选状态 |
| String getText() | 获取复选框文本信息 |
| void setText(String s) | 修改复选框文本信息 |

## 六、JRadioButton（单选按钮）

通常采用 ButtonGroup 将多个 JRadioButton 组成单选按钮组，实现多选一的功能。

### 1. JRadioButton 类的常用构造方法

（1）JRadioButton()：创建一个无文本的单选按钮。

（2）JRadioButton(String s)：创建一个指定显示文本的单选按钮。

（3）JRadioButton(String s，boolean selected)：创建一个指定显示文本，并指定初始选定状态的单选按钮。

## 2. JRadioButton 类的常用方法

JRadioButton 类的常用方法如表 5 -1 -7 所示。

表 5 -1 -7　JRadioButton 类的常用方法

| 方法 | 功能说明 |
|---|---|
| boolean isSelected( ) | 判断单选按钮是否被单击 |
| void setSelected( Boolean b) | 设置单选按钮是否被单击状态 |
| String getText( ) | 获取单选按钮文本信息 |
| void setText( String s) | 修改单选按钮文本信息 |

**【例5 -1 -7】** 代码如下：

```
1  public class P5_1_11 extends JFrame{
2      //声明组件
3      JPanel p1,p2,p3;
4      JLabel lb1,lb2;
5      JCheckBox cb1,cb2,cb3;
6      JRadioButton rb1,rb2;
7      JButton b1,b2;
8      public P5_1_11(){
9          //创建组件
10         p1 = new JPanel();
11         p2 = new JPanel();
12         p3 = new JPanel();
13         lb1 = new JLabel("你喜欢的运动");
14         lb2 = new JLabel("你的性别");
15         cb1 = new JCheckBox("足球");
16         cb2 = new JCheckBox("篮球");
17         cb3 = new JCheckBox("排球");
18         rb1 = new JRadioButton("男");
19         rb2 = new JRadioButton("女");
20         b1 = new JButton("注册用户");
21         b2 = new JButton("取消用户");
22         //添加
23         p1.add(lb1);
24         p1.add(cb1);
25         p1.add(cb2);
26         p1.add(cb3);
27         p2.add(lb2);
28         p2.add(rb1);
29         p2.add(rb2);
30         p3.add(b1);
31         p3.add(b2);
32         ButtonGroup bg = new ButtonGroup();
```

```
33          bg.add(rb1);
34          bg.add(rb2);
35          //设置布局(网格布局)
36          GridLayout gl = new GridLayout(3,1);
37          this.setLayout(gl);
38          //添加面板
39          this.add(p1);
40          this.add(p2);
41          this.add(p3);
42          this.setSize(400, 300);
43          this.setDefaultCloseOperation(JFrame.EXIT_ON_CLOSE);
44          this.setVisible(true);
45      }
46      public static void main(String[] args) {
47          P5_1_11  p = new P5_1_11      ();
48      }
49}
```

运行结果如图 5-1-7 所示。

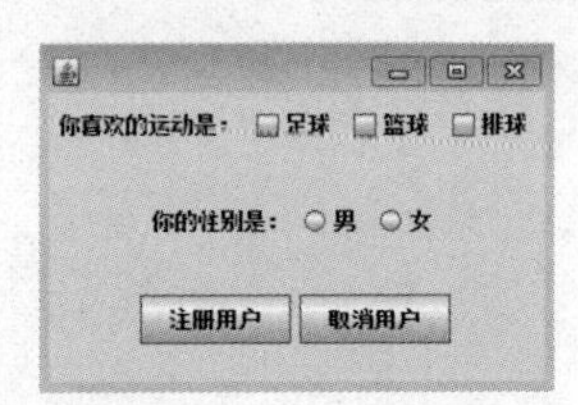


图 5-1-7　运行结果

## 七、JComboBox（组合框）

JComboBox 允许用户选择一个或者多个选项，默认下该组件隐藏所有列表项，当用户单击时则以下拉列表的形式显示所有列表项。

### 1. JComboBox 类的常用构造方法

（1）JComboBox()：创建一个空的组合框。

（2）JComboBox(Object[] items)：创建一个包含指定数组元素的组合框。

### 2. JComboBox 类的常用方法

JComboBox 类的常用方法如表 5-1-8 所示。

表 5-1-8　JComboBox 类的常用方法

| 方法 | 功能说明 |
| --- | --- |
| int getSelectIndex() | 获取当前选定项的索引 |
| Object getSelectItem() | 获取当前选定项 |
| void addItem(Object obj) | 在组合框列表项中添加新选项 |
| void insertItemAt(Object obj, int index) | 在组合框列表项指定位置插入新选项 |

续表

| 方法 | 功能说明 |
| --- | --- |
| Object getItemAt( int index) | 获取指定索引处的列表项 |
| void removeAllItems( ) | 删除所有选项 |

## 八、JList（列表框）

JList 允许用户从列表项中选择一个或多个。

### 1. JList 类的常用构造方法

（1）JList( )：创建一个空的列表框。

（2）JList( Object [ ] listData)：创建一个包含指定数组元素的列表框。

### 2. JList 类的常用方法

JList 类的常用方法如表 5 -1 -9 所示。

表 5 -1 -9　JList 类的常用方法

| 方法 | 功能说明 |
| --- | --- |
| int getSelectIndex( ) | 获取当前选定项的索引 |
| Object getSelectValue( ) | 获取第一个选定项 |
| Object[ ] getSelectValues( ) | 获取当前所有选定项 |
| Boolean isSelectdIndex( int index) | 返回列表框中是否有选定项 |

【例5 -1 -8】代码如下：

```
1  public class P5_1_12 {
2      public static void main(String[ ] args) {
3          JFrame frame = new JFrame( );
4          //设置窗体布局为网格布局,分为上、下两个网格
5          GridLayout gl = new GridLayout(2,1);
6          frame.setLayout(gl);
7          //创建两个面板,分别填充窗体中的两个网格
8          JPanel p1 = new JPanel( );
9          JPanel p2 = new JPanel( );
10         JComboBox combox = new JComboBox( ); //创建组合框
11         //添加组合框选项
12         combox.addItem("男");
13         combox.addItem("女");
14         JLabel lb1 = new JLabel("请选择性别");
15         //完成性别面板的设置
16         p1.add(lb1);
17         p1.add(combox);
18         //创建姓名数组
19         String boy[ ] = {"王强","李明","丁明","周扒皮","陈飞扬"};
```

```
20          //创建列表框,列表项为姓名数组
21          JList list = new JList(boy);
22          //设置列表框显示的列表项数量
23          list.setVisibleRowCount(3);
24          //为列表框添加滚动条
25          JScrollPane sp = new JScrollPane(list);
26          //完成姓名面板的设置
27          JLabel lb2 = new JLabel("请选择姓名");
28          p2.add(lb2);
29          p2.add(sp);
30          //将两个面板添加到窗体中
31          frame.add(p1);
32          frame.add(p2);
33          frame.setTitle("组合框、列表框");
34          frame.setSize(250,200);
35          frame.setVisible(true);
36          frame.setDefaultCloseOperation(JFrame.EXIT_ON_CLOSE);
37      }
38  }
```

**案例**

学生信息管理系统主要对学生信息进行管理，创建其录入学生信息界面。

**1. 录入学生信息界面的组成**

录入学生信息界面包括输入学号、姓名、性别、出生日期、专业、电话、地址信息等组件。

**2. 录入学生信息界面的设计**

1）组件选择

（1）标签：学号、姓名、性别、出生日期、专业、电话、地址。

（2）文本框：学号、姓名、出生日期、电话、地址。

（3）下拉组合框：专业。

（4）单选按钮 + 按钮组：性别。

（5）按钮：提交。

代码如下：

```
private JLabel lb1,lb2,lb3,lb4,lb5,lb6,lb7;
private JTextField tf1,tf2,tf3,tf4,tf5;
private JRadioButton rb1,rb2;
private JComboBox cb1;
private JPanel jp1,jp2,jp3,jp4,jp5,jp6,jp7,jp8;
private JButton jb1;
GridLayout gl = new GridLayout(8,1);
this.setLayout(gl);
```

2）布局选择

（1）窗体：网格布局（分为 8 个网格）。

（2）每个网格添加一个面板：流式分局（默认）。

3）创建窗体

代码如下：

```
public class Input extends JFrame{
    private JLabel lb1,lb2,lb3,lb4,lb5,lb6,lb7;
    private JTextField tf1,tf2,tf3,tf4,tf5;
    private JRadioButton rb1,rb2;
    private JComboBox cb1;
    private JPanel jp1,jp2,jp3,jp4,jp5,jp6,jp7,jp8;
    private JButton jb1;
    public Input() {
        lb1 = new JLabel("学号:");
        lb2 = new JLabel("姓名:");
        lb3 = new JLabel("性别:");
        lb4 = new JLabel("出生日期:");
        lb5 = new JLabel("专业:");
        lb6 = new JLabel("电话:");
        lb7 = new JLabel("地址:");
        tf1 = new JTextField(10);
        tf2 = new JTextField(10);
        tf3 = new JTextField(10);
        tf4 = new JTextField(10);
        tf5 = new JTextField(10);
        rb1 = new JRadioButton("男");
        rb2 = new JRadioButton("女");
        ButtonGroup bg = new ButtonGroup();
        bg.add(rb1);
        bg.add(rb2);
        cb1 = new JComboBox(new String[] {"物联网","网络","软件"});
        jb1 = new JButton("提交");
        GridLayout gl = new GridLayout(8,1);
        this.setLayout(gl);
        jp1 = new JPanel();
        jp2 = new JPanel();
        jp3 = new JPanel();
        jp4 = new JPanel();
        jp5 = new JPanel();
        jp6 = new JPanel();
        jp7 = new JPanel();
        jp8 = new JPanel();
        jp1.add(lb1);jp1.add(tf1);
        jp2.add(lb2);jp2.add(tf2);
        jp3.add(lb3);jp3.add(rb1);jp3.add(rb2);
        jp4.add(lb4);jp4.add(tf3);
        jp5.add(lb5);jp5.add(cb1);
        jp6.add(lb6);jp6.add(tf4);
        jp7.add(lb7);jp7.add(tf5);
        jp8.add(jb1);
        this.add(jp1);
        this.add(jp2);
        this.add(jp3);
        this.add(jp4);
```

```
        this.add(jp5);
        this.add(jp6);
        this.add(jp7);
        this.add(jp8);
        this.setSize(400, 500);
        this.setTitle("录入学生信息");
        this.setVisible(true);
        this.setDefaultCloseOperation(JFrame.EXIT_ON_CLOSE);
    }
    public static void main(String[] args) {
        Input input = new Input();
    }
}
```

运行结果结果如图 5 -1 -8 所示。

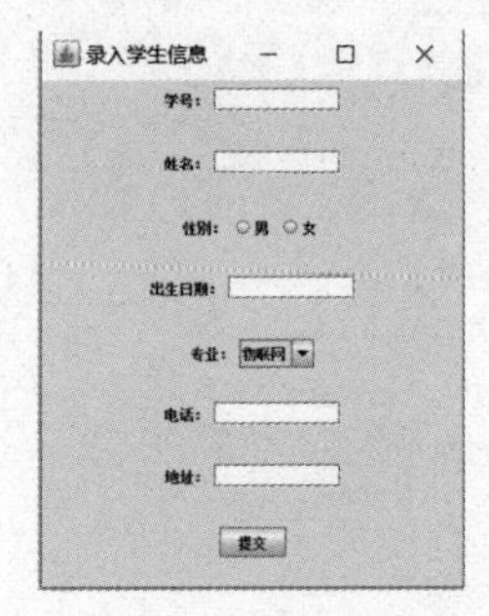


图 5 -1 -8　运行结果

## 任务工单

<table>
<tr><td>任务名称</td><td colspan="5">图书信息管理系统——图书查询窗体设计</td></tr>
<tr><td>班级</td><td></td><td>学号</td><td></td><td>姓名</td><td></td></tr>
<tr><td>任务要求</td><td colspan="5">（1）设计精确查询图书信息窗体。<br>（2）设计模糊查询图书信息窗体。</td></tr>
<tr><td>任务实现</td><td colspan="5"></td></tr>
<tr><td>异常及待解决问题记录</td><td colspan="5"></td></tr>
</table>

## 5.2 事件处理

### 学习目标

（1）掌握 GUI 中的事件处理机制。

（2）掌握事件处理方法。

### 相关知识

### 5.2.1 事件模型

Java 的事件处理机制主要包含 3 个事件处理对象：事件源、事件、事件监听器。下面以求两个文本框之和为例，借助“监考”案例，理解 3 个事件处理对象，如表 5-2-1 所示。

表 5-2-1 3 个事件处理对象

| 事件处理对象 | 监考 | 两个文本框求和 |
| --- | --- | --- |
| 事件（一个动作） | “作弊”行为 | 鼠标单击动作 |
| 事件源（监视动作发生在谁身上） | 学生 | “求和”按钮 |
| 事件监听器<br>（监视动作的发生及发生后的处理方法） | 监考教师 | 实现监听接口的类 |

为了让组件接收并处理用户事件，首先必须为组件注册相应的事件处理程序，即事件监听器，它是实现了对应事件监听程序接口的一个类。当事件产生时，将产生一个事件对象，事件监听器可以根据产生的事件对象来决定处理事件的方法。例如，为了处理命令按钮上的 ActionEvent 事件，需要定义一个实现 ActionListener 接口的事件监听程序类。通过 public void addActionListener（ActionListener l）方法为按钮组件注册事件监听程序，该方法的参数是一个实现 ActionListener 接口的类的实例，该实例既实现了事件监听，又实现了事件处理。图 5-2-1 所示为事件处理过程。

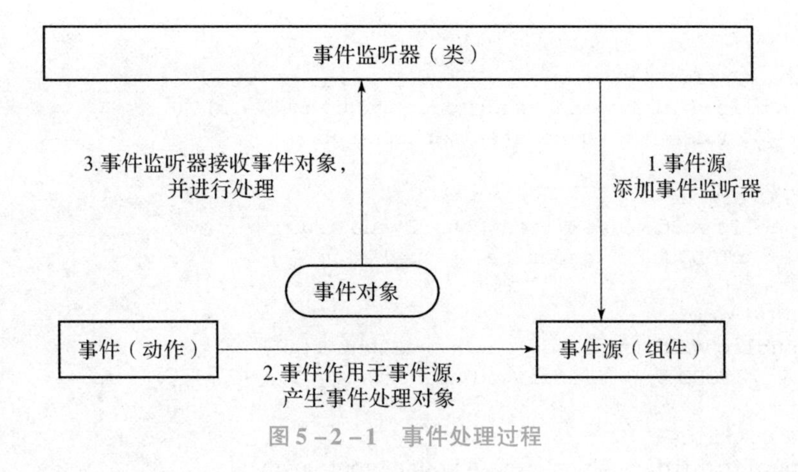


图 5－2－1　事件处理过程

### 5.2.2　事件模型的实现方法

事件模型的实现方法有 3 种：定义类实现监听接口、定义内部类实现监听接口和定义匿名内部类实现监听接口。

## 一、定义类实现监听接口

【例 5－2－1】代码如下：

```
1  //创建类,实现监听接口并成为一个事件监听器
2  public class P5_2_1 extends JFrame implements MouseListener{
3      private JButton b1,b2;
4      public P5_2_1 () {
5          FlowLayout fl = new FlowLayout();//布局管理器
6          this .setLayout(fl);
7          //创建组件,添加组件到窗体
8          b1 = new JButton("关闭窗体");
9          b2 = new JButton("设置窗体的标题");
10         this .add(b1);
11         this .add(b2);
12         b1.addMouseListener(this ); //为按钮注册事件监听器
13         this .setSize(400, 300);
14         this .setVisible(true );
15         this .setDefaultCloseOperation(JFrame. EXIT_ON_CLOSE );
16     }
17     public static void main(String[] args) {
18         P5_2_1 l = new P5_2_1();//创建窗体
19     }
20     //实现监听接口中的所有抽象方法
21     @ Override
22     public void mouseClicked(MouseEvent arg0) {
23         System.exit(0); //鼠标单击后相应的事件处理代码
```

```
24         }
25         @ Override
26         public void mouseEntered(MouseEvent arg0) {
27             //TODO Auto - generated method stub
28         }
29         @ Override
30         public void mouseExited(MouseEvent arg0) {
31             //TODO Auto - generated method stub
32         }
33         @ Override
34         public void mousePressed(MouseEvent arg0) {
35             //TODO Auto - generated method stub
36         }
37         @ Override
38         public void mouseReleased(MouseEvent arg0) {
39             //TODO Auto - generated method stub
40         }
41     }
```

运行结果如图 5－2－2 所示。

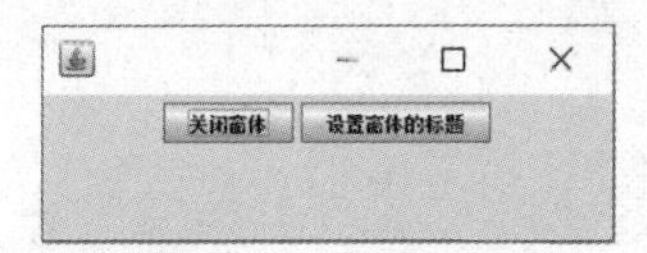


图 5－2－2 运行结果

## 二、内部定义类实现监听接口

【例 5－2－2】代码如下：

```
1  public class P5_2_2 extends JFrame{
2      private JButton b1;
3      public P5_2_2 () {
4          FlowLayout fl = new FlowLayout();
5          this.setLayout(fl);
6          b1 = new JButton("关闭窗体");
7          this.add(b1);
8          b1.addActionListener(new Inner());//为按钮注册事件监听器
9          this.setSize(400, 300);
10         this.setVisible(true);
11         this.setDefaultCloseOperation(JFrame.EXIT_ON_CLOSE);
12     }
13     //定义内部类,使监听接口成为事件监听器
14     public class Inner implements ActionListener{ //Inner 类成为一个事件监听器
15         @ Override
16         public void actionPerformed(ActionEvent e) {
17             System.exit(0);//按钮被单击后相应的事件处理代码
18         }
19     }
20     public static void main(String[] args) {
21         P5_2_2 p = new P5_2_2 ();
```

```
22      }
23  }
```

## 三、定义匿名内部类实现监听接口

**【例5-2-3】** 代码如下：

```
1  public class P5_2_3 extends JFrame{
2      private JButton b1;
3      public P5_2_3() {
4          FlowLayout fl = new FlowLayout();
5          this .setLayout(fl);
6          b1 = new JButton("关闭窗体");
7          this .add(b1);
8          //注册事件监听器,类实现了监听接口,只是这个实现类没有名字,为匿名类
9          b1.addActionListener(new ActionListener() {
10             @ Override
11             public void actionPerformed(ActionEvent e) {
12                 System.exit(0);
13             }
14         });
15         this .setSize(400, 300);
16         this .setVisible(true );
17         this .setDefaultCloseOperation(JFrame. EXIT_ON_CLOSE );
18     }
19     public static void main(String[ ] args) {
20         P5_2_3 p = new P5_2_3();
21     }
22 }
```

### 5.2.3　事件类、监听接口、事件适配器类

不同事件源上发生的事件种类不同，不同的事件由不同的监听者处理。java.awt.event包中定义了多种事件类。每个事件类都有一个对应的监听接口，监听接口中声明了若干抽象的事件处理方法，事件监听程序类需要实现相应的监听接口。

## 一、事件类

AWT事件分为两大类：低级事件和高级事件。

**1. 低级事件（基于组件的事件）**

（1）ComponentEvent组件事件：组件的大小、位置的变化等。

（2）ContainerEvent容器事件：容器中组件的添加、删除等。

（3）WindowEvent窗口事件：窗口的打开、关闭、最大化、最小化等。

（4）FocusEvent焦点事件：焦点的获得、失去等。

（5）KeyEvent键盘事件：键盘按钮的按下、释放等。

（6）MouseEvent鼠标事件：鼠标按键的按下、释放，鼠标的拖放等。

**2. 高级事件（基于语义的事件）**

（1）ActionEvent 动作事件：按钮被按下、在文本框中按 Enter 键被按下等。

（2）AdjustmentEvent 调整事件：数值的改变、滑块的移动等。

（3）ItemEvent 选项事件：选项的选定、取消等。

（4）TextEvent 文本事件：文本对象的改变等。

java. util. EventObject 类是所有事件类的基础父类，所有事件都是由它派生的。事件类的体系结构如图 5－2－3 所示。

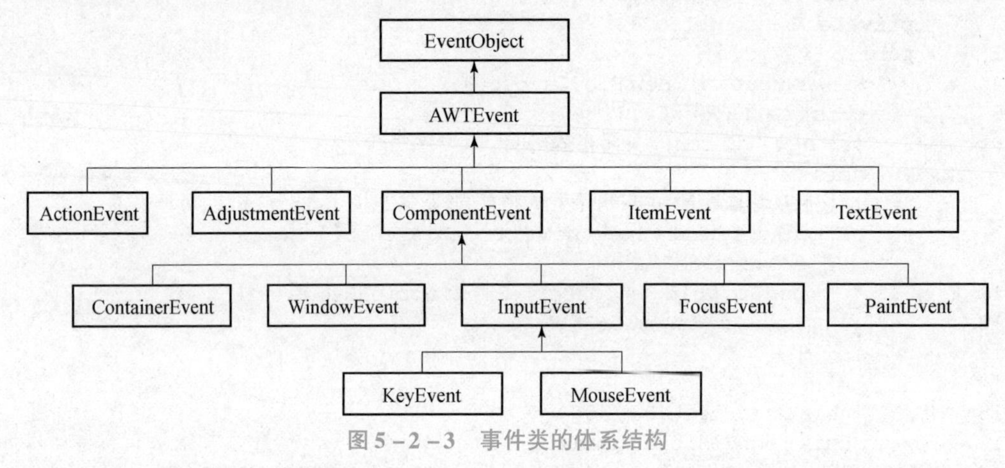


图 5－2－3　事件类的体系结构

## 二、监听接口

每个事件都有相应的监听接口，监听接口定义了与该事件有关的方法，每个方法都有一个用来接收事件源发送过来的事件对象，事件监听器可以从事件对象中获取事件的相关信息。Java 为 AWT 组件类和 Swing 组件类提供注册和注销事件监听器的方法，注册事件监听器的方法为 public void add × ×Listener（listener），如果不需要对该事件进行监听处理，可以把事件源的事件监听器注销，方法为 public void remove × ×Listener（listener）。表 5－2－2 所示为常用的 AWT 事件及其相应的监听接口。

表 5－2－2　常用的 AWT 事件及其相应的监听接口

| 事件类 | 描述信息 | 监听接口名 | 监听接口方法 |
| --- | --- | --- | --- |
| ActionEvent | 激活组件 | ActionListener | actionPerformed( ActionEvent e) |
| ItemEvent | 选择某些项目 | ItemListener | itemStateChanged( ItemEvent e) |
| MouseEvent | 鼠标移动 | MouseMotionListener | mouseDragged( MouseEvent e)<br>mouseMoved( MouseEvent e) |
| | 鼠标单击等 | MouseListener | mouseClicked( MouseEvent e)<br>mouseEntered( MouseEvent e)<br>mouseExited( MouseEvent e)<br>mousePressed( MouseEvent e)<br>mouseReleased( MouseEvent e) |

续表

| 事件类 | 描述信息 | 监听接口名 | 监听接口方法 |
| --- | --- | --- | --- |
| KeyEvent | 键盘输入 | KeyListener | keyPressed(KeyEvent e)<br>keyReleased(KeyEvent e)<br>keyTyped(KeyEvent e) |
| FocusEvent | 组件收到或失去焦点 | FocusListener | focusGained(FocusEvent e)<br>focusLost(FocusEvent e) |
| AdjustmentEvent | 移动滚动条等组件 | AdjustmentListener | adjustmentValueChanged(AdjustmentEvent e) |
| ComponentEvent | 对象移动、缩放、显示、隐藏等 | ComponentListener | componentHidden(ComponentEvent e)<br>componentShown(ComponentEvent e)<br>componentMoved(ComponentEvent e)<br>componentResized(ComponentEvent e) |
| WindowEvent | 窗口收到窗口级事件 | WindowListener | windowOpened(WindowEvent e)<br>windowClosing(WindowEvent e)<br>windowClosed(WindowEvent e)<br>windowActivated(WindowEvent e)<br>windowDeactivated(WindowEvent e)<br>windowIconified(WindowEvent e)<br>windowDeiconified(WindowEvent e) |
| ContainerEvent | 在容器中增加、删除组件 | ContainerListener | componentAdded(ContainerEvent e)<br>componentRemoved(ContainerEvent e) |
| TextEvent | 文本字段或文本区发生改变 | TextListener | textValueChanged(TextEvent e) |

### 1. 鼠标事件及其监听接口

在 GUI 中，鼠标主要用来选择、切换或绘画。所有组件都可以产生鼠标事件，通过实现 MouseListener 接口和 MouseMotionListener 接口来处理相应的鼠标事件。

（1）MouseListener 接口，主要针对鼠标的按键及坐标进行检测，共提供如下 5 个事件处理方法。

①public void mouseClicked (MouseEvent e)：鼠标单击事件。

②public void mouseEntered (MouseEvent e)：鼠标进入事件。

③public void mouseExited(MouseEvent e)：鼠标离开事件。

④public void mousePressed (MouseEvent e)：鼠标按下事件。

⑤public void mouseReleased(MouseEvent e)：鼠标松开事件。

（2）MouseMotionListener 接口，主要针对鼠标的拖动操作进行处理，有如下 2 个事件处理方法。

①public void mouseMoved(MouseEvent e)：鼠标移动事件。

②publiv void mouseDragged(MouseEvent e)：鼠标拖动事件。

（3）MouseEvent 类还提供了获取发生鼠标事件的坐标及单击次数的成员方法。MouseEvent 类中的常用方法如下。

①Point getPoint( )：返回 Point 对象，包含鼠标事件发生的坐标点。

②int getClickCount( )：返回与此鼠标事件关联的鼠标的单击次数。

③int getX( )：返回鼠标事件 X 坐标。

④int getY( )：返回鼠标事件 Y 坐标。

⑤int getButton( )：返回哪个鼠标按键更改了状态。

**【例5-2-4】** 代码如下：

```
1  public class P5_2_4 extends JFrame{
2      public P5_2_4 () {
3          //窗体添加鼠标事件监听器,实现鼠标监听接口中的5个抽象方法
4          this.addMouseListener(new MouseListener() {
5          @ Override
6          public void mouseClicked(MouseEvent e) {
7              System. out .println("鼠标被单击了,当前位置为:
                                   x = " + e.getX() + ",y = " + e.getY());
8          }
9          @ Override
10          public void mouseExited(MouseEvent e) {
11              System. out .println("鼠标移出窗体");
12          }
13          @ Override
14          public void mouseEntered(MouseEvent e) {
15              System. out .println("鼠标进入窗体");
16          }
17          @ Override
18          public void mousePressed(MouseEvent e) {
19              System. out .println("鼠标按键被按下了");
20          }
21          @ Override
22          public void mouseReleased(MouseEvent e) {
23              System. out .println("鼠标按键被松开了");
24          }
25      });
26      this.setSize(300,200);
27      this.setVisible(true);
28      this.setDefaultCloseOperation(JFrame. EXIT_ON_CLOSE);
29    }
30    public static void main(String[] args) {
31        P5_2_4 p = new P5_2_4 ();
32    }
33  }
```

运行结果如图 5-2-4 所示。

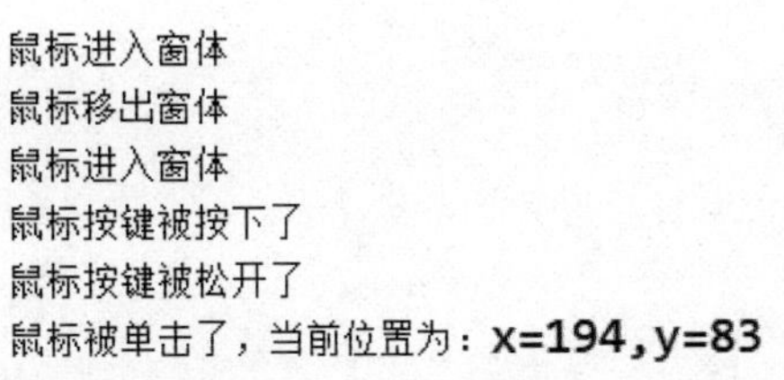
鼠标进入窗体
鼠标移出窗体
鼠标进入窗体
鼠标按键被按下了
鼠标按键被松开了
鼠标被单击了，当前位置为：x=194,y=83

图5-2-4 运行结果

【说明】

程序中第4行为窗体添加了鼠标事件监听器，必须实现鼠标监听接口中的所有抽象方法。从程序的运行结果来看，鼠标的5个动作均被监听到，并执行了相应的代码。

### 2. 键盘事件及其监听接口

键盘事件也是最常用的用户交互方式。Java 提供了 KeyEvent 类来捕获键盘事件，通过实现 KeyListener 接口来处理相应的键盘事件。

KeyListener 接口中有如下3个事件处理方法。

（1）public void keyPressed(KeyEvent e)：当键盘上某个键被按下时触发该方法，用于监听键的按下操作。

（2）public void keyReleased(KeyEvent e)：当键盘上某个键被释放时触发该方法，用于监听键的释放操作。

（3）public void keyTyped(KeyEvent e)：当键盘上某个键的字符被入时触发该方法，用于监听键盘的字符输入操作。

KeyEvent 类的常用方法如下。

（1）char getKeyChar( )：获取引发键盘事件的键对应的 Unicode 字符。若键没有相应的 Unicode 字符，则返回 KeyEvent 类的一个静态常量 KeyEvent. CHAR - UNDEFINED。

（2）String getKeyText( )：获取引发键盘事件的键的文本内容。

（3）int getKeyCode( )：获取与键盘事件中的键关联的整数 keyCode。

**【例5-2-5】**代码如下：

```
1  public class P5_2_5 extends JFrame{
2    private JPanel jp1,jp2,jp3;
3    private JLabel lb1,lb2,lb3;
4    private JTextField textField,textField_1,textField_2;
5    public P5_2_5 () {
6        this.setLayout(new GridLayout(3,1));
7        jp1 = new JPanel();
8        jp2 = new JPanel();
9        jp3 = new JPanel();
10       lb1 = new JLabel("输入 n1:");
11       lb2 = new JLabel("输入 n2:");
12       lb3 = new JLabel("累加和:");
13       textField = new JTextField();
14       textField_1 = new JTextField();
15       textField_2 = new JTextField();
16       jp1.add(lb1);
```

```
17        jp1.add(textField);
18        jp2.add(lb2);
19        jp2.add(textField_1);
20        jp3.add(lb3);
21        jp3.add(textField_2);
22        this.add(jp1);
23        this.add(jp2);
24        this.add(jp3);
25        //为第 2 个文本框添加键盘事件监听器,实现监听接口中的 3 个抽象方法
26        textField_1.addKeyListener(new KeyListener() {
27        @ Override//实现键被按下方法
28        public void keyPressed(KeyEvent arg0) {
29           //判断按下的键是不是 Enter 键,如果是,则开始计算
30           if (arg0.getKeyCode() == KeyEvent. VK_ENTER ) {
31              int n1 = Integer.parseInt(textField.getText());
32              int n2 = Integer.parseInt(textField_1.getText());
33              //保证 n1 < n2
34              int n;
35              if (n1 > n2) {
36                 n = n1;
37                 n1 = n2;
38                 n2 = n;
39              }
40             //求累加和
41             Integer sum = 0;
42             for (int i = n1;i <= n2;i ++)
43                sum = sum + i;
44             //展示结果
45             textField_2.setText(sum.toString());
46           }
47        }
48        //实现键被松开方法
49        @ Override
50        public void keyReleased (KeyEvent arg0) {
51        }
52        //实现键的字符被输入方法
53        @ Override
54        public void keyTyped (KeyEvent arg0) {
55        }
56  });
57  this.setSize(300,200);
58  this.setVisible(true );
59  this.setDefaultCloseOperation(JFrame. EXIT_ON_CLOSE );
60  public static void main(String[] args) {
61      P5 -2 -5 p = new P5 -2 -5 ();
62  }
63 }
```

运行结果如图 5-2-5 所示。

图 5 -2 -5　运行结果

【说明】

程序第 26 行为第 2 个文本框添加了键盘事件监听器，第 28、50、54 行分别实现了键盘监听接口中的 keyPressed( )、keyReleased( )、keyTyped( )方法。

## 三、事件适配器类

从上述例子可以看出，事件处理需要实现监听接口，这些监听接口中往往声明了多个抽象方法，为了实现事件监听，则需要实现监听接口中的所有抽象方法。例如 WindowListener 接口中定义了 7 个抽象方法，在实现监听接口的类中必须同时实现这 7 个抽象方法。然而，用户往往只关心其中的某个或者某几个抽象方法。为了简化编程，引入事件适配器（Adapter）类。Java 为具有两个以上方法的监听接口配备了一个 × × Adapter 类，提供了监听接口中所有抽象方法的缺省实现。开发中，将不再直接实现监听接口，只需继承事件适配器类并重写需要的事件处理方法即可。表 5 -2 -3所示为常用事件适配器类。

表 5 -2 -3　常用事件适配器类

| 事件类 | 监听接口名 | 事件适配器类 |
| --- | --- | --- |
| ActionEvent | ActionListener | 只包含一个抽象方法，不提供事件适配器 |
| ItemEvent | ItemListener | 只包含一个抽象方法，不提供事件适配器 |
| MouseEvent | MouseMotionListener | MouseMotionAdapter |
| | MouseListener | MouseAdapter |
| KeyEvent | KeyListener | KeyAdapter |
| FocusEvent | FocusListener | FocusAdapter |
| AdjustmentEvent | AdjustmentListener | 只包含一个抽象方法，不提供事件适配器 |
| ComponentEvent | ComponentListener | ComponentAdapter |
| WindowEvent | WindowListener | WindowAdapter |
| ContainerEvent | ContainerListener | ContainerAdapter |
| TextEvent | TextListener | 只包含一个抽象方法，不提供事件适配器 |

**【例 5 -2 -6】**代码如下：

```
1  public class P5_2_6 extends JFrame{
2      private JPanel jp1,jp2,jp3;
3      private JLabel lb1,lb2,lb3;
4      private JTextField textField,textField_1,textField_2;
5      public P5_2_6 () {
6          this.setLayout(new GridLayout(3,1));
7          jp1 = new JPanel();
8          jp2 = new JPanel();
9          jp3 = new JPanel();
10         lb1 = new JLabel("输入 n1:");
11         lb2 = new JLabel("输入 n2:");
12         lb3 = new JLabel("累加和:");
13         textField = new JTextField();
14         textField_1 = new JTextField();
15         textField_2 = new JTextField();
16         jp1.add(lb1);
17         jp1.add(textField);
18         jp2.add(lb2);
19         jp2.add(textField_1);
20         jp3.add(lb3);
21         jp3.add(textField_2);
22         this.add(jp1);
23         this.add(jp2);
24         this.add(jp3);
25         //为第 2 个文本框添加键盘监听器,通过事件适配器类仅实现 keyPressed()方法
26         textField_1.addKeyListener(new KeyAdapter() {
27           @ Override
28           public void keyPressed(KeyEvent arg0) {
29             //判断按下的键是不是 Enter 键,如果是,则开始计算
30             if (arg0.getKeyCode() == KeyEvent. VK_ENTER ) {
31                 //获取两个数
32                 int n1 = Integer.parseInt(textField.getText());
33                 int n2 = Integer.parseInt(textField_1.getText());
34                 //保证 n1 < n2
35                 int n;
36                 if (n1 > n2) {
37                     n = n1;
38                     n1 = n2;
39                     n2 = n;
40             }
41             //求累加和
42             Integer sum = 0;
43             for (int i = n1;i < = n2;i ++)
44                 sum = sum + i;
45             textField_2.setText(sum.toString());
46             }
47           }
48         });
49        this.setSize(300,200);
```

```
50            this.setVisible(true);
51            this.setDefaultCloseOperation(JFrame.EXIT_ON_CLOSE);
52            public static void main(String[] args) {
53                P5_2_6 p = new P5_2_6();
54            }
55        }
```

运行结果如图 5 -2 -6 所示。

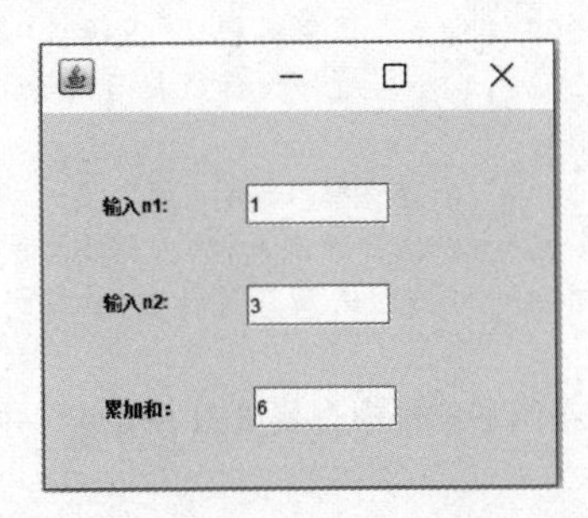


图 5 -2 -6　运行结果

【说明】

程序第 26 行通过事件适配器类 KeyAdapter 创建键盘监听器，只需要复写 keyPressed( ) 方法，其他两个无须使用方法，可以忽略。

## 任务工单

<table>
<tr><td>任务名称</td><td colspan="5">图书信息管理系统——图书信息精确查询功能</td></tr>
<tr><td>班级</td><td></td><td>学号</td><td></td><td>姓名</td><td></td></tr>
<tr><td>任务要求</td><td colspan="5">（1）定义图书集合，代码如下：<br><br>List<Book> books = new ArrayList<Book>();<br>books.add(new Book("001","Java 程序设计","王强","清华大学出版社",50,"在库"));<br>books.add(new Book("002","Java 案例教程","王新","北京大学出版社",58,"在库"));<br>books.add(new Book("003","Java 程序设计从入门到精通","周北平","清华大学出版社",46,"在库"));<br>books.add(new Book("004","C 语言程序设计","何贝","高等教育出版社",38,"在库"));<br>books.add(new Book("005","C 语言案例教程","周华金","高等教育出版社",40,"在库"));<br>books.add(new Book("006","Python 程序设计","林琳","北京大学出版社",50,"在库"));<br><br>（2）根据图书编号，在图书集合中查找该图书信息，并展示。</td></tr>
<tr><td>任务实现</td><td colspan="5">(提交交互代码)</td></tr>
<tr><td>异常及待解决问题记录</td><td colspan="5"></td></tr>
</table>

# 5.3 WindowBuilder 的安装及使用

## 学习目标

（1）掌握 WindowBuilder 插件的安装方法。
（2）掌握使用 WindowBuilder 修改布局方式、创建组件的方法。
（3）掌握使用 WindowBuilder 实现事件处理的方法。

## 相关知识

### 5.3.1 WindowBuilder 的安装

#### 一、下载 WindowBuilder 安装包（WindowBuilder – Updates –1.13.0.zip）

下载地址为 https：// eclipse. dev/windowbuilder/download. php。
具体实现步骤如图 5 –3 –1、图 5 –3 –2 所示。

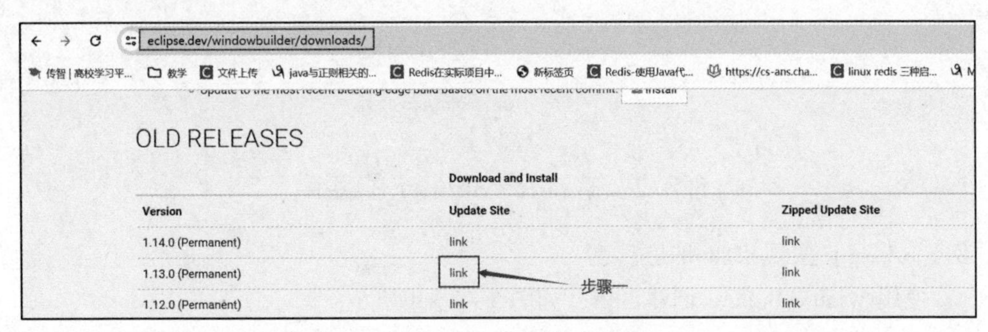


图 5 –3 –1　步骤（1）

图 5 –3 –2　步骤（2）

#### 二、安装 WindowBuilder

在“Help”菜单中选择“Install New Software”命令，进入图 5 –3 –3 所示界面，选择

安装包路径，选择安装所有插件，安装完毕后重新启动 Eclipse 即可。

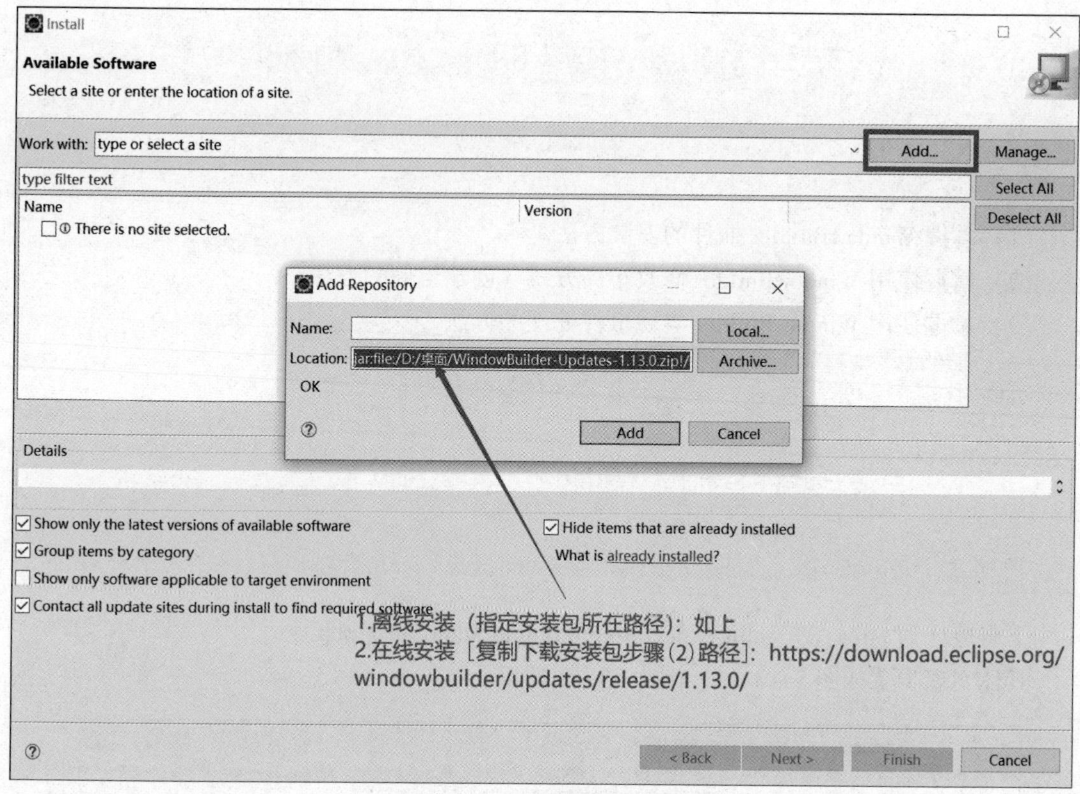


图 5－3－3　选择安装包路径

## 5.3.2　WindowBuilder 的使用

**【例 5－3－1】** 实现累加和功能。

（1）使用 WindowBuilder 创建窗体，如图 5－3－4 所示。

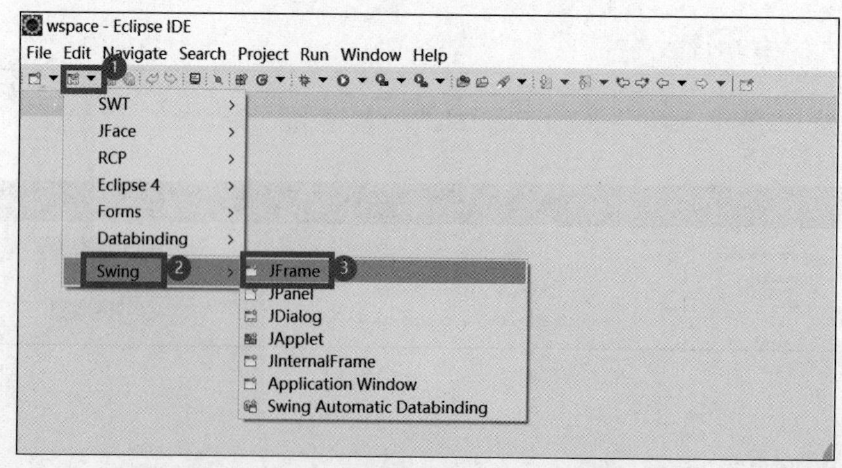


图 5－3－4　使用 WindowBuilder 创建窗体

将“Source”选项卡切换到“Design”选项卡，进入图形设计界面，如图 5－3－5 所示。

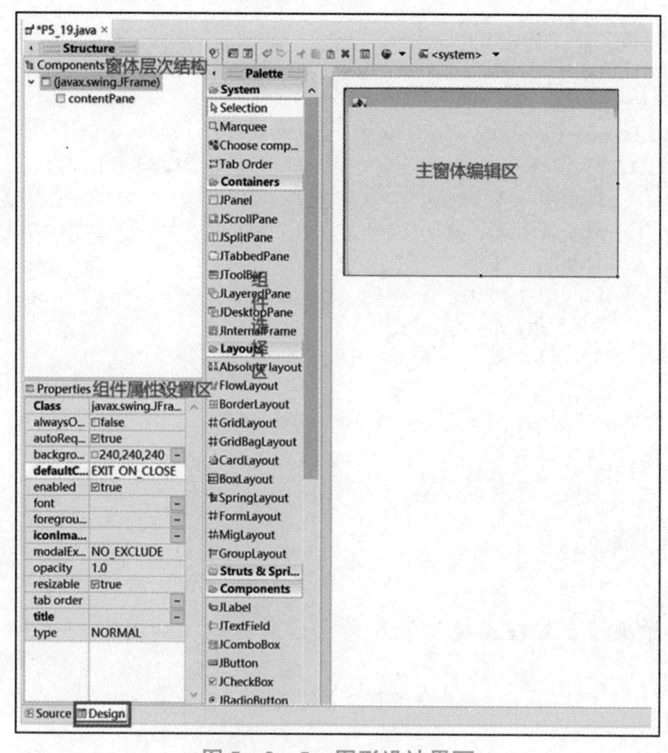


图 5－3－5　图形设计界面

（2）使用 WindowBuilder 添加事件监听器。

选择事件源，通过鼠标右键快捷菜单添加键盘监听器，如图 5－3－6 所示。

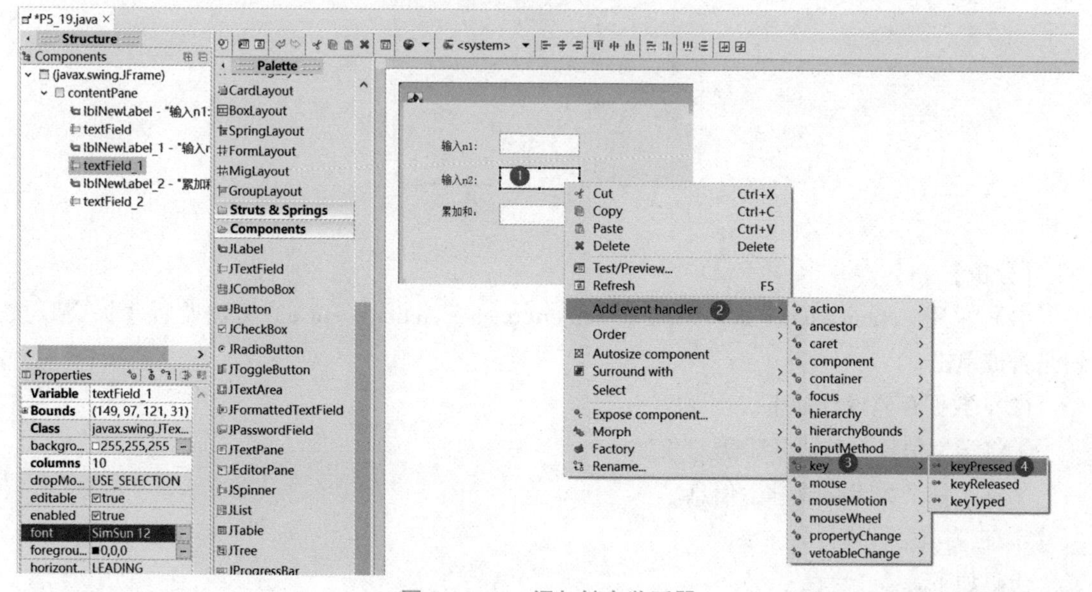


图 5－3－6　添加键盘监听器

切换至“Source”选项卡，系统已经自动添加匿名类监听器，并实现监听功能，代码如下。

```
1      textField_1 = new JTextField();
2      textField_1.addKeyListener(new KeyAdapter() {
3          @ Override
4          public void keyPressed(KeyEvent e) {
5              if (e.getKeyCode() == KeyEvent. VK_ENTER ) {
6                  int n1 = Integer.parseInt(textField.getText());
7                  int n2 = Integer.parseInt(textField_1.getText());
8                  int n;
9                  if (n1 >n2) { n = n1; n1 = n2; n2 = n; }
10                  Integer sum = 0;
11                  for (int i = n1;i < = n2;i ++)
12                      sum = sum + i;
13                  textField_2.setText(sum.toString());
14              }
15          }
16      });
```

案例

实现 5.1 节中学生信息管理系统的录入学生信息功能，如图 5-3-7 所示。

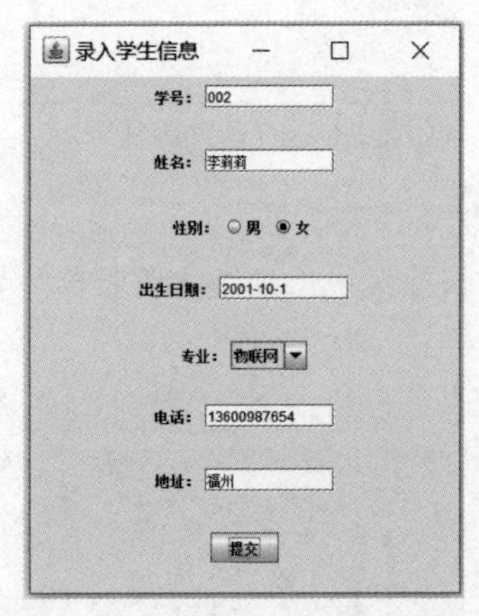


图 5-3-7　录入学生信息界面

【分析】

(1) 采用 ActionListener 接口中的 actionPerformed（ActionEvent e）方法监听“提交”按钮是否被单击。

(2) 数据存储设计如下。

①将学生信息封装成学生类对象。

②定义一个数组存放学生信息。

**1. 定义学生封装类**

代码如下：

```
public class Student {
    private String s_id, s_name,s_sex;
    private Date s_birthday;
    private String s_major;
    private int s_scholarship;
    private String s_photo, s_qrcode;
    //访问方法、设置方法
    //构造方法
    //toString()方法
}
```

**2. 数据存储准备**

代码如下：

```
//定义存放学生信息的数组
Student[] stus = new Student[10];
```

**3. 实现学生信息录入功能**

代码如下：

```
i = 0; //数组下标
jb1 = new JButton("提交");
jb1.addActionListener(new ActionListener() {
    public void actionPerformed(ActionEvent e) {
        Student s = new Student();
        s.setS_id(tf1.getText());
        s.setS_name(tf2.getText());
        if (rb1.isSelected())
            s.setS_sex("男");
        else
            s.setS_sex("女");
        SimpleDateFormat format = new SimpleDateFormat("yyyy-MM-dd");
        Date date = null ;
        try {
            date = format.parse(tf3.getText());
        } catch (ParseException e1) {
          //TODO Auto-generated catch block
          e1.printStackTrace();
        }
        s.setS_birthday(date);
        s.setS_major(cb1.getSelectedItem().toString());
        s.setTel(tf4.getText());
        s.setAddress(tf5.getText());
        stus[i] = s;
        i ++;
    }
});
```

**任务工单**

| 任务名称 | 图书信息管理系统——图书信息管理增、删、改窗体 | | | | |
|---|---|---|---|---|---|
| 班级 | | 学号 | | 姓名 | |
| 任务要求 | （1）定义图书集合，代码如下：<br>List<Book> books = new ArrayList<Book>();<br>books.add(new Book("001","Java 程序设计","王强","清华大学出版社",50,"在库"));<br>books.add(new Book("002","Java 案例教程","王新","北京大学出版社",58,"在库"));<br>books.add(new Book("003","Java 程序设计从入门到精通","周北平","清华大学出版社",46,"在库"));<br>books.add(new Book("004","C 语言程序设计","何贝","高等教育出版社",38,"在库"));<br>books.add(new Book("005","C 语言案例教程","周华金","高等教育出版社",40,"在库"));<br>books.add(new Book("006","Python 程序设计","林琳","北京大学出版社",50,"在库"));<br>（2）根据图书编号，在图书集合中查找该图书信息，并展示。 | | | | |
| 任务实现 | （提交窗体截图及交互代码） | | | | |
| 异常及待解决问题记录 | | | | | |

## 练习题

### 1. 选择题

（1）下列组件中（　　）不是 Component 的子类。

A. Button　　B. Dialog　　C. Label　　D. MenuBar

（2）在 AWT 中，常用的布局管理器包括（　　）。（多选）

A. FlowLayout 布局管理器　　B. BorderLayout 布局管理器

C. CardLayout 布局管理器　　D. GridLayout 布局管理器

### 2. 填空题

（1）大部分 Swing 组件是__________类的直接或者间接子类，其名称都是在原来 AWT 组件的名称前加字母 J。

（2）每个容器都有一个默认的布局管理器，如果不希望通过布局管理器对容器进行布局，可以调用容器的__________方法将其取消。

### 3. 设计题

（1）创建题图 5－1 所示界面。

题图 5－1　设计题（1）图

（2）设计题图 5－2 所示界面。

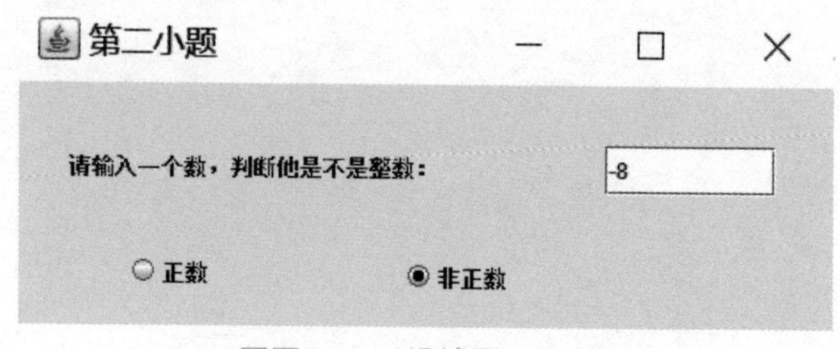


题图 5－2　设计题（2）图

# 第6章

# 集合类

——学如逆水行舟，不进则退

# 6.1 List 接口

## 学习目标

（1）理解集合的概念。

（2）了解 Collection 接口、List 接口。

（3）掌握 List 接口的实现类 ArrayList、LinkedList 的应用开发。

（4）掌握集合的遍历。

## 相关知识

### 6.1.1 集合概述

在 Java 程序中，集合是一个专门用来存储 Java 类对象的容器。在前面的章节中，使用数组保存一组对象，但在实际程序开发过程中，有时很难预先确定需要保存对象的个数，此时数组将不再适用，因为数组的长度是不可变的。例如，新生报到时，要保存所有报到新生的信息，由于不确定即将来报到的新生的数量，所以无法确定数组的长度。为了在程序中可以保存这些数目不确定的对象，JDK 提供了一系列特殊的类，这些类可以存储任意类型的对象，并且长度可变，在 Java 中这些类统称为集合。集合都存在于 java. util 包中，在使用时一定要注意导包的问题，否则会出现异常。

集合按照其存储结构可以分为两大类，即单列集合和双列集合，这两种集合的特点如下。

（1）单列集合。单列集合用于存储一系列符合某种规则的元素，它的父接口是 Collection 接口，它有两个重要的子接口，分别是 List 接口和 Set 接口。其中，List 接口的特点是元素有序、可重复；Set 接口的特点是元素无序，而且不可重复。List 接口的主要实现类有 ArrayList 和 LinkedList，Set 接口的主要实现类有 HashSet 和 TreeSet。

（2）双列集合。双列集合用于存储具有键（Key）、值（Value）映射关系的元素，每个元素都包含一对键值，在使用双列集合时可以通过指定的键找到对应的值。例如，在双列集合中键保存学号，值保存学生对象，则根据学生的学号就可以找到对应的学生对象。双列接口的主要实现类有 HashMap 和 TreeMap，它的父接口是 Map 接口。

图 6－1－1 所示为集合的继承体系，其中白底框是接口类，灰底框是具体的实现类。

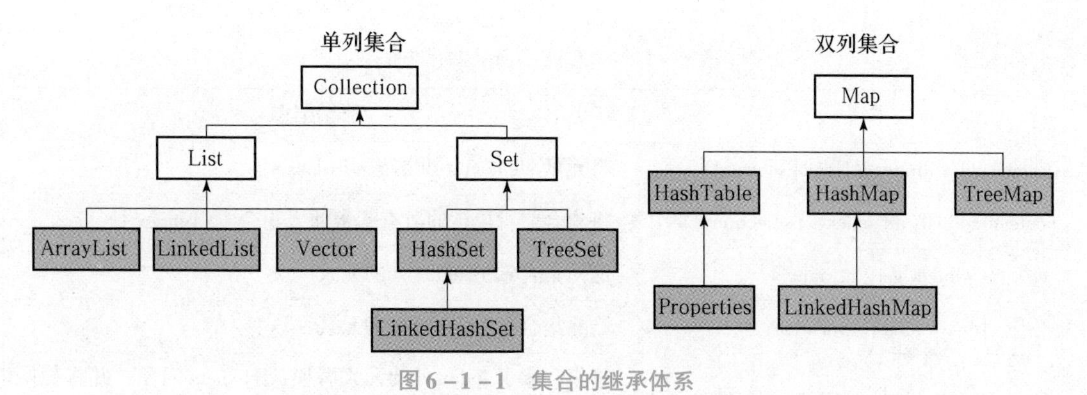


图 6－1－1　集合的继承体系

### 6.1.2　Collection 接口

Collection 接口是单列集合的父接口。Collection 接口的常用方法如表 6－1－1 所示。

表 6－1－1　Collection 接口的常用方法

| 方法 | 功能描述 |
| --- | --- |
| boolean add(Object o) | 向集合中添加一个元素 |
| boolean addAll(Collection c) | 将指定 Collection 接口中的所有元素添加到该集合中 |
| void clear() | 删除该集合中的所有元素 |
| boolean remove(Object o) | 删除该集合中的指定元素 |
| boolean removeAll(Collection c) | 删除指定集合中的所有元素 |
| boolean isEmpty() | 判断该集合是否为空 |
| Boolean contains(Object o) | 判断该集合是否包含某个元素 |
| boolean containsAll(Collection c) | 判断该集合是否包含指定集合中的所有元素 |
| Iterator iterator() | 返回在该集合的元素上进行迭代的迭代器（Iterator），用于遍历该集合的所有元素 |
| int size() | 获取该集合中元素的个数 |

### 6.1.3　List 接口

List 接口继承 Collection 接口，是单列集合的一个重要分支，通常将实现了 List 接口的对象称为 List 集合。List 集合具有如下特点：

（1）允许出现重复的元素，集合中的所有元素以一种线性的方式进行存储，访问集合中的指定元素可通过索引定位。

（2）元素是有序的，即元素的存入顺序和取出顺序一致。

List 接口作为 Collection 接口的子接口，继承了 Collection 接口的全部方法，并且增加了一些根据元素索引来操作集合的特有方法，如表 6－1－2 所示。

表 6-1-2　List 接口的常用方法

| 方法 | 功能描述 |
| --- | --- |
| void add(intindex, Object element) | 将元素 element 入集合索引 index 处 |
| boolean addAll(int index, Collection c) | 将集合 c 所包含的所有元素插入集合索引 index 处 |
| Object get(int index) | 返回集合索引 index 处的元素 |
| Object remove(tnt index) | 删除集合索引 index 处的元素 |
| Object set(int index, Object element) | 将集合索引 index 处的元素替换成 element 对象，并将替换后的元素返回对象。 |
| int indexOf(Object o) | 返回在 List 集合中出现的位置索引 |
| int IastIndexOf(Object o) | 返回在 List 集合中最后一次出现的位置索引 |
| List subList(int fromlndex, int tolndex) | 返回从索引 from fromlndex（包括）到 tolndex（不包括）处的所有元素 |

### 6.1.4　ArrayList 集合

ArrayList 集合是 List 接口的一个实现类，它是使用频率最高的一种集合。可以把 ArrayList 集合理解为一个长度可变的数组，系统会在内存中为每个 ArrayList 集合元素分配一个存储空间，数量随程序变化。由于 ArrayList 集合的底层是使用一个数组来保存元素的，在指定位置增加或删除元素会导致创建新的数组，效率比较低，所以 ArrayList 集合不适用于大量的增删操作，但其数组结构允许程序通过索引的方式访问元素，因此使用 ArrayList 集合查找元素很便捷。

#### 一、ArrayList 集合的定义

定义 ArrayList 集合的语法格式如下：

```
List[ <泛型 >] 集合名 = new ArrayList[ <泛型 >]();
```

或

```
ArrayList [ <泛型 >] 集合名 = new ArrayList[ <泛型 >]();
```

（1）使用 ArrayList 集合必须使用 import java. util. ArrayList 导包，否则系统会报参数类型错误。在程序开发中，往往要用到多个集合。因此，可以使用“import java. util. *”语句，利用通配符“*”，一次性导入包中的所有类。

（2）“<泛型>”可省略。泛型用来指定集合元素的数据类型。若未声明泛型，则系统默认 ArrayList 集合元素的数据类型为 Object，可以在 ArrayList 集合中存入任意数据类型的元素，因为 Object 类型是所有类的直接父类或间接父类。若已声明泛型，则 ArrayList 集合元

素必须严格使用泛型所指定的数据类型。例如：

```
List s = new ArrayList();
s.add("ddd");
s.add(1);
s.add('c');
List < Student > stuList = new ArrayList < Student >();
stuList.add("ddd");//报参数类型错误
```

## 二、ArrayList 集合的使用

ArrayList 集合实现了 List 接口，List 接口的父接口是 Collection 接口，因此 ArrayList 集合继承了 Collection 接口及 List 接口的所有方法。其中，最常用的是 add()方法、remove()方法、set()方法、get()方法，用于实现元素的添加、删除、存值、取值。

**【例 6-1-1】** 利用 ArrayList 集合解决新生报到案例中新生信息存储的问题。在前面的章节中，使用数组存放新生信息，数组的长度由录取学生数量决定。为了不造成存储空间的浪费，采用 ArrayList 集合动态地为新生分配内存单元。代码如下：

```
1  import java.util.ArrayList;
2  public class Student {
3      private String sid,name,sex;
4      private int age;
5      private String dormId, institute; //宿舍、所在学院
6      //省略 get()、set()方法、构造方法、toString()方法
7      public static void main(String[] args) {
8         //定义集合 stuList,存放所有报到新生信息
9         ArrayList < Student > stuList = new ArrayList < Student >();
10         //新生 1
11         Student s1 = new Student("001", "林丽娟", "女", 19, "西 001", "外语");
12         //调用 add()方法,将新生 1 信息存储到集合 stuList 中
13         stuList.add(s1);
14         //调用 get()方法,获取集合中的第 1 个元素并输出
15         System.out.println("修改前,第 1 个集合元素:" + stuList.get(0));
16         //新生 2
17         Student s2 = new Student("002", "王鹏", "男", 19, null , "土建");
18         stuList.add(s2);
19         //调用 get()方法,获取集合中的第 2 个元素,并修改该元素的宿舍属性值
20         stuList.get(1).setDormId("东 001");
21         System.out.println("第 2 个集合元素:" + stuList.get(1));
22         Student s3 = new Student("001", "林小娟", "女", 19, "西 001", "外语");
23         //调用 set()方法,将集合中的第 1 个元素值更改为 s3
24         stuList.set(0, s3);
25         //调用 get()方法,将集合中的第 1 个元素输出
26         System.out.println("修改后,第 1 个集合元素:" + stuList.get(0));
27         System.out.println("集合的长度为:" + stuList.size());
28      }
29  }
```

运行结果如图 6-1-2 所示。

```
修改前,第1个集合元素：Student [sid=001, name=林丽娟, sex=女, age=19, dormId=西001, institute=外语]
第2个集合元素：Student [sid=002, name=王鹏, sex=男, age=19, dormId=东001, institute=土建]
修改后,第1个集合元素：Student [sid=001, name=林小娟, sex=女, age=19, dormId=西001, institute=外语]
集合的长度为：2
```

图 6-1-2 运行结果

【说明】

程序第 1 行，集合都存在于 java. util 包中，使用时要注意导入正确的包。第 9 行，新建一个用于存放 Student 类对象的集合 stuList，若存入其他类对象，则系统将给出错误提示。第 13 行，调用 add( )方法将新生对象 s1 添加到集合 stuList 中，此时集合长度为 1。第 15 行，调用 get(0)方法，获取集合 stuList 中索引值为 0 的元素的值，这里需要注意，集合的索引与数组是一样的，索引值为 0 表示集合中的第 1 个元素，索引的取值范围是从 0 开始的，最后一个索引值是 size-1，在访问元素时要注意索引不可超出范围，否则会抛出下标越界异常 IndexOutOfBoundsException。第 24 行，调用 set( )方法，将集合中索引值为 0 的元素值改为对象 s3。第 27 行，调用 size( )方法，输出集合长度。

### 6.1.5 LinkedList 集合

ArrayList 集合的存储结构适合查询操作多的应用，而对于增删操作较多的情况，Java 提供了 List 接口的另一个实现类 LinkedList。该集合内部是一个双向循环链表，链表中的每个元素都使用引用的方式来记住它的前一个元素和后一个元素，从而可以将所有元素彼此连接起来。当插入一个新元素时，只需要修改元素之间的这种引用关系即可，删除一个元素时也是如此。正因为具有这样的存储结构，LinkedList 集合对于元素的增删操作有很高的效率。

图 6-1-3 所示为新增元素示意。元素 1 和元素 2 在原集合中互为前后关系，在它们之间增一个新元素，只需要将新元素设置为元素 1 的下一个元素、元素 2 的上一个元素即可。

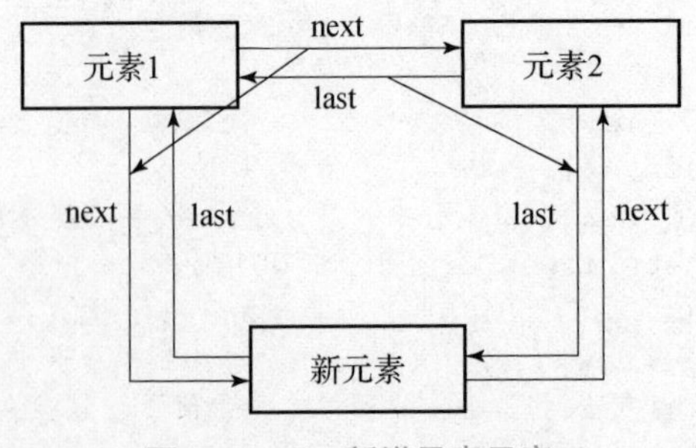


图 6-1-3 新增元素示意

图 6-1-4 所示为删除元素示意。原来集合中，待删除元素介于元素 1 和元素 2 之间，现将待删除元素删除，只需要将元素 1 的下一个元素设置为元素 2，将元素 2 的上一个元素设置为元素 1 即可，即使元素 1 与元素 2 互为前后关系。

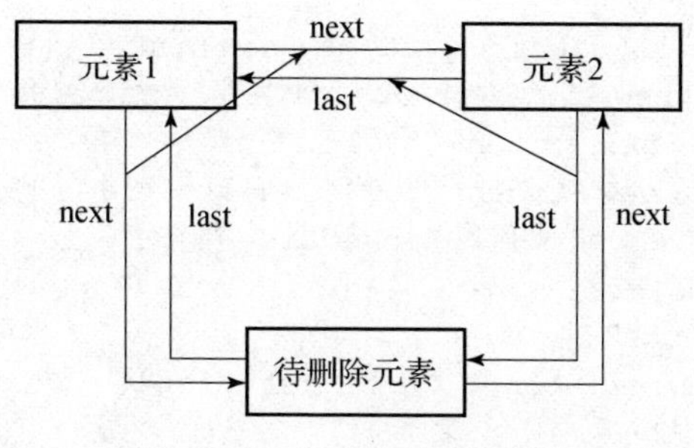


图6－1－4　删除元素示意

## 一、LinkedList 集合的定义

定义 LinkedList 集合的语法格式如下：

```
List[<泛型>] 集合名 = new LinkedList[<泛型>]();
```

或

```
LinkedList [<泛型>] 集合名 = new LinkedList [<泛型>]();
```

## 二、LinkedList 集合的使用

LinkedList 集合具有增删元素效率高的特点，为此，LinkedList 集合中定义了实现增删操作的方法，如表6－1－3所示。

表6－1－3　LinkedList 集合的常用方法

| 方法 | 功能描述 |
|---|---|
| void add(int index, E element) | 在列表中指定的位置插入指定的元素 |
| void addFirst(Object o) | 将指定元素插入列表的开头 |
| void addLast(Object o) | 将指定元素添加到列表的结尾 |
| Object getFirst() | 返回列表的第一个元素 |
| Object getLast() | 返回列表的最后一个元素 |
| Object removeFirst() | 移除并返回列表的第一个元素 |
| Object removeLast() | 移除并返回列表的最后一个元素 |

**【例6－1－2】** 代码如下：

```
1  import java.util.ArrayList;
2  public class P6_1_2 {
3      public static void main(String[] args) {
```

```
4         LinkedList<Student> stuList = new LinkedList<Student>();
5         Student s1 = new Student("001", "林丽娟", "女", 19, "西001", "外语");
6         Student s2 = new Student("002", "王鹏", "男", 19, null, "土建");
7         Student s3 = new Student("003", "陈萌", "男", 18, "东003", "信息");
8         //将学生对象s1,s2存储到LinkedList集合中
9         stuList.add(s1);
10        stuList.add(s2);
11        System.out.println("插入前" + stuList);
12        //将对象s3插入到s1、s2之间
13        stuList.add(1, s3);
14        System.out.println("插入后" + stuList);
15        //删除LinkedList集合中第1个学生对象
16        stuList.remove(0);
17        stuList.removeLast();
18        System.out.println("删除后:" + stuList);
19    }
20 }
```

运行结果如图 6-1-5 所示。

```
插入前: [Student [sid=001, name=林丽娟, sex=女, age=19, dormId=西001, institute=外语], Student [sid=002, name=王鹏, sex=男, age=19, dormId=null, institute=土建]]
插入后: [Student [sid=001, name=林丽娟, sex=女, age=19, dormId=西001, institute=外语], Student [sid=003, name=陈萌, sex=男, age=18, dormId=东003, institute=信息], Student [sid=002, name=王鹏, sex=男, age=19, dormId=null, institute=土建]]
删除后: [Student [sid=003, name=陈萌, sex=男, age=18, dormId=东003, institute=信息]]
```

图 6-1-5　运行结果

### 6.1.6　Vector 集合

Vector 集合与 ArrayList 集合相似，都是基于数组实现的，但是 Vector 集合是线程安全的，ArrayList 集合是线程不安全的。Vector 集合采用同步锁，为访问它的方法提供了线程安全保障，主要用于多线程并发访问的场景；ArrayList 集合主要用于单线程环境下的读取、修改操作等。由于获取同步锁需要额外的开销，所以 Vector 集合的使用性能较 ArrayList 集合差。

#### 一、Vector 集合的定义

定义 Vector 集合的语法格式如下：

```
List[<泛型>] 集合名 = new Vector[<泛型>]();
```

或

```
Vector [<泛型>] 集合名 = new Vector [<泛型>]();
```

#### 二、Vector 集合的使用

Vector 集合的常用方法如表 6-1-4 所示。

表6-1-4 Vector集合的常用方法

| 常用方法 | 功能描述 |
|---|---|
| void add(int index, Element e) | 在列表中的指定位置插入指定元素 |
| void addAll(int index, Collection c) | 在指定位置插入集合c |
| Object remove(Object o) | 移除指定元素，并返回该元素 |
| Object get(int index) | 返回指定位置的元素 |
| int indexOf(Object o) | 返回指定元素的位置 |
| boolean contains(Object o) | 判断列表是否包含指定元素 |

**【例6-1-3】**代码如下：

```
1   import java.util.*;
2   public class P6_1_3 {
3       public static void main(String[] args) {
4           Vector<Student> vec = new Vector<Student>();
5           Student s1 = new Student("001", "林丽娟", "女", 19, "西001", "外语");
6           Student s2 = new Student("002", "王鹏", "男", 19, null, "土建");
7           vec.add(s1);
8           vec.add(s2);
9           for (Student student : vec) {
10              System.out.println(student);
11          }
12          System.out.println(vec.contains(s1));
13          System.out.println(vec.indexOf(s2));
14          System.out.println(vec.get(0));
15      }
16  }
```

运行结果如图6-1-6所示。

```
Student [sid=001, name=林丽娟, sex=女, age=19, dormId=西001, institute=外语]
Student [sid=002, name=王鹏, sex=男, age=19, dormId=null, institute=土建]
true
1
Student [sid=001, name=林丽娟, sex=女, age=19, dormId=西001, institute=外语]
```

图6-1-6 运行结果

### 6.1.7 迭代器

在程序开发中，除了使用集合来保存对象，还经常需要遍历集合中的所有元素实现对象的查找及更新等操作。JDK专门提供了一个遍历集合的接口Iterator，它与Collection接口、Map接口不同，Collection接口与Map接口主要用于存储元素，而Iterator接口用于迭代访问（遍历）Collection接口中的元素，因此Iterator对象也称为迭代器。

## 一、迭代器的定义

定义迭代器的语法格式如下：

```
Iterator<泛型> 迭代器名 = 集合.iterator();
```

定义迭代器时，如果没有指定集合元素数据类型，则迭代器返回 Object 对象，否则返回具体数据类型。例如：

```
Iterator ite1 = stuList.iterator();
Object obj = ite1.next();//迭代器返回对象为 Object 类型
Iterator<Student> ite2 = stuList.iterator();
Student s = ite2.next();//迭代器返回对象为 Student 类型
```

## 二、迭代器的使用

使用迭代器的语法格式如下：

```
while(迭代器.hasNext()){
    泛型  对象名 = 迭代器.next();
        .....
}
```

使用 next()方法获取元素时必须配合 hasNext()方法，以保证要获取的元素存在，否则会抛出 NoSuchElementException 异常。

**【例 6-1-4】** 以新生报到案例学习如何使用迭代器遍历集合中的元素。代码如下：

```
1  import java.util.ArrayList;
2  public class P6_1_4 {
3      public static void main(String[] args) {
4          LinkedList<Student> stuList = new LinkedList<Student>();
5          Student s1 = new Student("001", "林丽娟", "女", 19, "西 001", "外语");
6          Student s2 = new Student("002", "王鹏", "男", 19, null , "土建");
7          Student s3 = new Student("003", "陈萌", "男", 18, "东 003", "信息");
8          //将学生对象 s1,s2 存储到 LinkedList 集合中
9          stuList.add(s1);
10         stuList.add(s2);
11         stuList.add(s3);
12         //需求 1. 输出所有新生信息
13         Iterator<Student> ite = stuList.iterator();
14         while (ite.hasNext()) {
15             s = ite.next();
16             System. out .println(s);
17         }
18         //需求 2. 统计外语学院报到人数
19         Iterator ite1 = stuList.iterator();
```

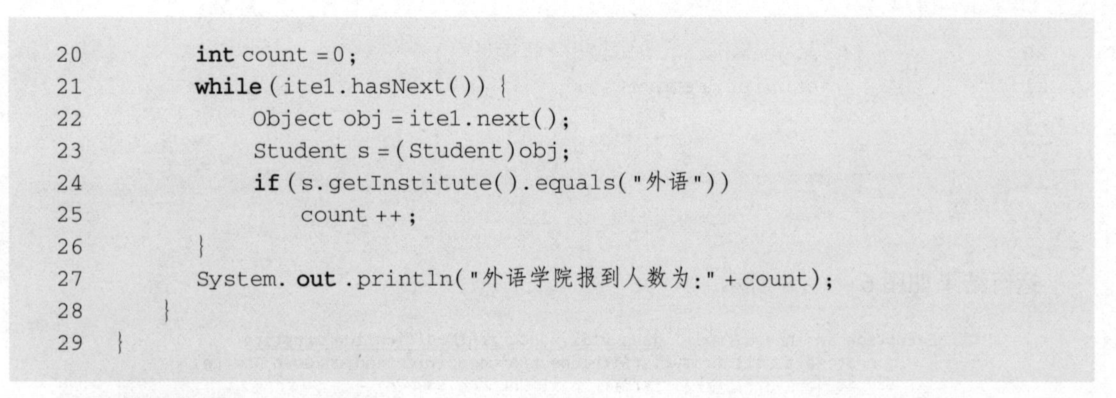

```
20          int count =0;
21          while (ite1.hasNext()) {
22              Object obj = ite1.next();
23              Student s = (Student)obj;
24              if (s.getInstitute().equals("外语"))
25                  count ++;
26          }
27          System.out.println("外语学院报到人数为:" + count);
28      }
29  }
```

运行结果如图 6－1－7 所示。

```
Student [sid=001, name=林丽娟, sex=女, age=19, dormId=西001, institute=外语]
Student [sid=002, name=王鹏, sex=男, age=19, dormId=null, institute=土建]
Student [sid=003, name=陈萌, sex=男, age=18, dormId=东003, institute=信息]
外语学院报到人数为：1
```

图 6－1－7　运行结果

【说明】

首先，第 13 行通过调用 stuList 集合的 iterator( ) 方法获取迭代器对象；然后，使用 hasNext( ) 方法判断集合中是否存在下一个元素，如果存在，则调用 next( ) 方法将元素取出，否则说明已到达集合末尾，停止遍历元素。

## 三、迭代器的异常删除

### 1. ConcurrentModificationException 异常

使用迭代器对集合中的元素进行遍历时，如果调用了集合对象的 remove( ) 方法删除元素之后继续使用迭代器遍历元素，则会出现异常。下面通过【例 6－1－5】演示这种异常。

**【例 6－1－5】**以新生报到案例为例，在所有已报到的新生中，突然有一位新生决定回高三复读，这时就需要在遍历集合时找出这位新生并将其删除。

```
7   public class P6_1_5 {
8       public static void main(String[] args) {
9           List < Student > stuList = new ArrayList < Student >();
10          Student s1 = new Student("001", "林丽娟", "女", 19, "西 001", "外语");
11          Student s2 = new Student("002", "王鹏", "男", 19, null , "土建");
12          Student s3 = new Student("003", "陈萌", "男", 18, "东 003", "信息");
13          stuList.add(s1);
14          stuList.add(s2);
15          stuList.add(s3);
16          //在集合中删除姓名为"林丽娟"的元素
17          Iterator < Student > ite = stuList.iterator();
18          while (ite.hasNext()) {
19              Student s = ite.next();
```

```
20            if(s.getName().equals("林丽娟")) {
21                stuList.remove(s);
22            }
23        }
24    }
25 }
```

运行结果如图 6-1-8 所示。

```
Exception in thread "main" java.util.ConcurrentModificationException
	at java.util.ArrayList$Itr.checkForComodification(Unknown Source)
	at java.util.ArrayList$Itr.next(Unknown Source)
	at chapter07.Test.main(Test.java:20)
```

图 6-1-8　运行结果

【说明】

Java 使用迭代器遍历 List 对象，事实上是将对 List 对象的遍历托管给迭代器，所有对 List 对象进行的增删操作都必须经过迭代器，否则迭代器遍历时会出错，因此这里直接对 List 对象的元素进行删除时，迭代器就抛出 ConcurrentModificationException 异常。出错的根本原因是迭代器在遍历前已经从 List 对象处获取需要的迭代次数，遍历条件 hasNext（）就是根据这个迭代次数来判断是否已经迭代完毕；而在遍历过程中，任意删除 List 对象的元素会导致迭代器的迭代次数与 List 对象的元素个数不一致。

**2. 两种异常解决方法**

方法 1：在 List 对象中将要删除的元素删除后，则跳出循环不再继续迭代。例如下面的代码，补充第 16 行的 break 语句即可。

```
11        Iterator<Student> ite = stuList.iterator();
12        while(ite.hasNext()) {
13            Student s = ite.next();
14            if(s.getName().equals("林丽娟")) {
15                stuList.remove(s);
16                break;
17            }
```

方法 2：使用迭代器本身的删除方法来删除元素。例如下面的代码，将第 15 行代码改为 ite. remove()即可。

```
11        Iterator<Student> ite = stuList.iterator();
12        while(ite.hasNext()) {
13            Student s = ite.next();
14            if(s.getName().equals("林丽娟")) {
15                ite.remove();
16            }
```

### 6.1.8　foreach 循环

在 Java 中，使用迭代器遍历集合中的元素，写法比较烦琐。因此，从 JDK 5.0 开始提

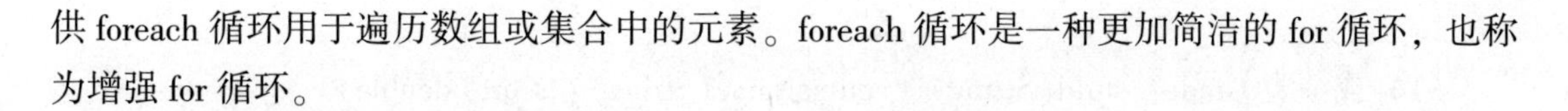

供 foreach 循环用于遍历数组或集合中的元素。foreach 循环是一种更加简洁的 for 循环，也称为增强 for 循环。

## 一、foreach 循环的语法格式

foreach 循环的语法格式如下：

```
for(容器中元素类型  循环对象:容器变量){
    语句体
}
```

## 二、foreach 循环的集合遍历

foreach 循环在遍历集合时语法简洁，没有循环条件，也没有迭代语句，所有这些工作都交给 Java 虚拟机完成。foreach 循环的次数是由容器中元素的个数决定的，每次循环时，foreach 循环都通过变量获取当前循环的元素。

**【例 6-1-6】** 代码如下：

```
1  public class P6_1_6 {
2      public static void main(String[] args) {
3          List<Student> stuList = new ArrayList<Student>();
4          Student s1 = new Student("001", "林丽娟", "女", 19, "西 001", "外语");
5          Student s2 = new Student("002", "王鹏", "男", 19, null, "土建");
6          Student s3 = new Student("003", "陈萌", "男", 18, "东 003", "信息");
7          stuList.add(s1);
8          stuList.add(s2);
9          stuList.add(s3);
10         for (Student s : stuList) {
11             s.setName("李白");
12             s = s3;//将 s 指向对象 s3, stuList 集合的元素并没有发生变化
13         }
14         System.out.println("采用 foreach 更新:" + stuList);
15     }
16 }
```

运行结果如图 6-1-9 所示。

```
采用foreach更新: [Student [sid=001, name=李白, sex=女, age=19, dormId=西001, institute=外语], Student [sid=002, name=李白,
sex=男, age=19, dormId=null, institute=土建], Student [sid=003, name=李白, sex=男, age=18, dormId=东003, institute=信息]]
```

图 6-1-9　运行结果

### 案例

在教务管理系统中，学生成绩管理模块的主要功能为录入学生成绩、修改学生成绩、删除学生成绩、查询学生成绩。那么如何实现学生成绩的存储？学习第 2 章后，可以知道用数组能够存储学生成绩。然而，学生成绩信息与日俱增，那么存储学生成绩的数组长度如何定义才合适？过量地定义数组长度将造成空间的浪费，小量地定义数组长度又担心不够用。有没有不用定义数组长度来解决这个问题的方法？

**1. 类的设计**

（1）成绩类 Grade：stuId(String)、courseName(String)、score(double)。

（2）学生成绩管理窗体：ListGradeFrame. java。

**2. 集合的设计**

（1）gradeList 集合：用于存放学生成绩，使用 ArrayList 集合。

（2）unpassList 集合：用于存放不及格的学生成绩，使用 LinkedList 集合。

**3. 窗体设计**

窗体设计如图 6-1-10 所示。

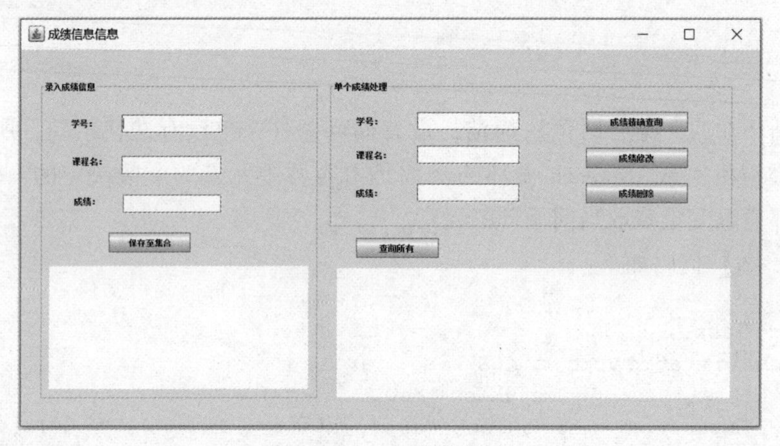


图 6-1-10 窗体设计

**4. 类的定义**

代码如下：

```
public class Grade {
    private String stuId , courseName;
    private double score;
}
```

**5. 功能实现**

（1）录入成绩信息，保存至集合，代码如下：

```
List<Grade> gradeList = new ArrayList<Grade>();
JButton button_3 = new JButton("保存至集合");
button_3.addActionListener(new ActionListener() {
    public void actionPerformed(ActionEvent arg0) {
        String stuId = textField.getText();
        String courseName = textField_1.getText();
        Double score = Double.parseDouble(textField_2.getText());
        Grade g = new Grade();
        g.setStuId(stuId);
        g.setCourseName(courseName);
        g.setScore(score);
```

```
            gradeList.add(g);
            String st = "";
            for (Grade grade : gradeList) {
                st = st + grade + " \n";
            }
            textArea.setText(st);
        }
    });
```

（2）成绩精确查询，代码如下：

```
JButton button = new JButton("成绩精确查询");
button.addActionListener(new ActionListener() {
      public void actionPerformed(ActionEvent arg0) {
          String stuId = textField_3.getText();
          String courseName = textField_4.getText();
          Iterator <Grade> ite = gradeList.iterator();
          while (ite.hasNext()) {
              Grade grade = ite.next();
if (grade.getStuId().equals(stuId)&&grade.getCourseName().equals(courseName))
              textField_5.setText("" + grade.getScore());
          }
      }
  });
```

（3）成绩修改，代码如下：

```
JButton button_1 = new JButton("成绩修改");
button_1.addActionListener(new ActionListener() {
      public void actionPerformed(ActionEvent arg0) {
          String stuId = textField_3.getText();
          String courseName = textField_4.getText();
          Double score = Double.parseDouble(textField_5.getText());
          for (Grade grade : gradeList) {
if (grade.getStuId().equals(stuId)&&grade.getCourseName().equals(courseName))
              grade.setScore(score);
          }
      }
  });
```

（4）成绩删除，代码如下：

```
JButton button_2 = new JButton("成绩删除");
button_2.addActionListener(new ActionListener() {
      public void actionPerformed(ActionEvent arg0) {
          String stuId = textField_3.getText();
          String courseName = textField_4.getText();
          Double score = Double.parseDouble(textField_5.getText());
```

```
            Grade g = new Grade(stuId,courseName,score);
            for (Grade grade : gradeList) {
    if (grade.getStuId().equals(stuId)&&grade.getCourseName().equals(courseName)) {
                gradeList.remove(grade);
                break;
            }
        }
      }
  });
```

（5）查询所有，代码如下：

```
JButton button_4  = new JButton("查询所有");
button_4.addActionListener(new ActionListener() {
      public void actionPerformed(ActionEvent arg0) {
      String stuId = textField_3.getText();
      String courseName = textField_4.getText();
      Double score = Double.parseDouble(textField_5.getText());
      String st = "";
      Iterator<Grade> ite = gradeList.iterator();
      while (ite.hasNext()) {
          Grade grade = ite.next();
          st = st + grade + "\n";
      }
      textArea_1.setText(st);
    }
  });
```

<table>
<tr><td>任务名称</td><td colspan="5">图书信息管理系统——批量添加图书信息</td></tr>
<tr><td>班级</td><td></td><td>学号</td><td></td><td>姓名</td><td></td></tr>
<tr><td>任务要求</td><td colspan="5">将数组中的图书添加到集合中</td></tr>
<tr><td>任务实现</td><td colspan="5"><pre>public class Test {
public static void main(String[] args) {
Book[] bookArr = new Book[]{
new Book("001","Java 程序设计","王强","清华大学出版社",50,"在库"),
new Book("002","Java 案例教程","王新","北京大学出版社",58,"在库"),
new Book("003","Java 程序设计从入门到精通","周北平","清华大学出版社",
46,"在库"),
new Book("004","C 语言程序设计","何贝","高等教育出版社",38,"已借"),
new Book("005","C 语言案例教程","周华金","高等教育出版社",40,"已借"),
new Book("006","Python 程序设计","林琳","北京大学出版社",50,"已借")
};
//创建图书集合

    //将数组中的图书添加到图书集合中

    }
    }</pre></td></tr>
<tr><td>调试过程记录</td><td colspan="5"></td></tr>
</table>

# 6.2 Set 接口

## 学习目标

（1）掌握 Set 接口的实现类 HashSet、TreeSet 的应用开发。

（2）掌握集合的遍历。

## 相关知识

### 6.2.1 Set 接口

Set 接口与 List 接口一样，继承 Collection 接口。与 List 接口不同的是，Set 接口中的元素无序，并且不重复。Set 接口主要有两个实现类，分别是 HashSet 集合和 TreeSet 集合。其中，HashSet 集合是根据对象的哈希值来确定元素在集合中的存储位置，因此它具有良好的存取和查找性能。TreeSet 集合则是以二叉树的方式存储元素，它可以对集合中的元素进行排序。

### 6.2.2 HashSet 集合

HashSet 集合是 Set 接口的一个实现类，HashSet 集合的底层数据结构是哈希表，哈希表其实就是数组单向链表，这样的存储结构决定了它具有较高的查询效率。HashSet 集合进行数据存储时，主要完成如下步骤。

步骤一，调用 hashcode( )方法，计算当前存入集合的新元素的哈希值。

步骤二，根据哈希值计算新元素在集合中的索引位置。

步骤三，判断在集合中此索引位置上是否有元素，如果没有，则直接存储新元素，否则调用 equals( )方法判断原位置上的元素内容与新元素内容是否相同，若不同，则存储新元素，否则不存储新元素。

定义 HashSet 集合的语法格式如下：

```
Set[<泛型>] 集合名 = new HashSet[<泛型>]();
```

或

```
HashSet[<泛型>] 集合名 = new HashSet[<泛型>]();
```

**【例 6-2-1】** 以新生报到案例学习如何使用 HashSet 集合。代码如下：

```
1  import java.util.*;
2  public class P6_2_1 {
```

```
3      public static void main(String[] args) {
4          Set < Student > hset = new HashSet < Student >();
5          Student s1 = new Student("001", "林丽娟", "女", 19, "西001", "外语");
6          Student s2 = new Student("002", "王鹏", "男", 19, null , "土建");
7          Student s3 = new Student("003", "陈萌", "男", 18, "东003", "信息");
8          Student s4 = new Student("003", "陈萌", "男", 18, "东003", "信息");
9          hset.add(s1);
10         hset.add(s2);
11         hset.add(s3);
12         hset.add(s4);
13         Iterator < Student > ite = hset.iterator();
14         while (ite.hasNext()) {
15             Student s = ite.next();
16             System. out .println(s);
17         }
18     }
19 }
```

运行结果如图6-2-1所示。

```
Student [sid=003, name=陈萌, sex=男, age=18, dormId=东003, institute=信息]
Student [sid=001, name=林丽娟, sex=女, age=19, dormId=西001, institute=外语]
Student [sid=002, name=王鹏, sex=男, age=19, dormId=null, institute=土建]
Student [sid=003, name=陈萌, sex=男, age=18, dormId=东003, institute=信息]
```

图6-2-1　运行结果

【说明】

程序定义了一个存放Student类对象的Hashset集合，先后创建了4个Student类对象存入集合，迭代器遍历所有元素。从运行结果可以看出，输出元素的顺序与添加元素的顺序不一致，但内容相同的对象s3和s4均被添加到集合中。

**【例6-2-2】**修正Student类，重写hashCode()和equals()方法，保证元素存储的唯一性。代码如下：

```
1  import java.util.ArrayList;
2  public class Student {
3      private String sid,name,sex;
4      private int age;
5      private String dormId, institute; //宿舍、所在学院
6      @ Override//重写 hashCode()方法
7      public int hashCode() {
8          final int prime = 31;
9          int result = 1;
10         result = prime * result + ((sid == null ) ? 0 : sid.hashCode());
11         return result; //返回属性 sid 的哈希值
12     }
13     @ Override//重写 equals()方法
```

```
14        public boolean equals(Object obj) {
15            if (this == obj) //判断是否是同一个对象,如果是则返回 true
16                return true ;
17            if (obj == null ) //若参数为空则返回 false
18                return false ;
19            //判断参数对象是否为 Student 类对象,如果不是则返回 false
20            if (getClass()!= obj.getClass())
21                return false ;
22            Student other = (Student) obj;//将对象强制转换为 Student 类型
23            if (sid == null ) {
24            if (other.sid!= null ) //判断对象 sid 是否为空
25                return false ;//若集合中 sid 为空且对象中用户名不为空,则返回 false
26            } else if (!sid.equals(other.sid)) //判断参数对象 sid 与原对象 sid 是否
27                                               //相同,若不同则返回 false
28                return false ;
29            return true ;
30        }
31        //省略 get()、set()方法、构造方法、toString()方法
32    }
```

【说明】

程序第 7 ~ 12 行重写了 Student 类的 hashCode( )方法，生成属性 sid 的哈希值；第 14 ~ 30 行重写了 Student 类的 equals( )方法，判断新对象与集合中指定位置的对象是否相同，如果相同则返回 true，重复的 Student 对象即被删除。但是，若 HashSet 集合中存入的对象不是自定义对象，则无序复写 hashCode( )方法及 equals( )方法。

**案例**

实现携程网用户注册功能。用户以手机号码注册成为携程网用户，要求每个手机号码只能注册一个携程用户。在程序中，用 List 集合存储所有携程网用户信息，注册新用户时，必须得遍历集合中的每个用户信息以保证手机号码未被注册，比较烦琐，是否有简单的方式来验证手机号码是否已被注册过呢?

【分析】

(1) 为了保存用户信息，必须定义一个用户类，存放用户的手机号码、姓名、性别、出生日期等信息。

(2) 定义 HashSet 集合，存放用户信息，保证手机号码不被重复注册。

**1. 类的设计**

(1) 用户类 User：tel（String）、name（String）、sex（String）、birthday（Date）、password（String）。

(2) 携程网用户注册窗体：RegisterFrame. java。

**2. 集合的设计**

UserList 集合：用于存放用户信息，采用 HashSet 集合。

**3. 类的定义**

定义 User 类的代码如下：

```
import java.text.SimpleDateFormat;
import java.util.Date;
public class User {
      private String tel ,name ,sex;
      private Date birthday;
      private String password;
      //省略除了 birthday 之外其他变量的 get()、set()方法、构造方法、toString()方法
      public String getBirthday() {
          SimpleDateFormat format = new SimpleDateFormat("yyyy - MM - dd");
          String birth = format.format(birthday);
          return birth;
}
public void setBirthday(Date birthday) {
    this .birthday  = birthday;
}
@ Override //重写 hashCode()方法
public int hashCode() {
    final int prime  = 31;
    int result  = 1;
    result  = prime  * result  + ((tel  == null ) ? 0 : tel.hashCode());
    return result;
  }
  @ Override//重写 equals()方法
  public boolean equals(Object obj) {
      if (this  == obj) return true ;
      if (obj  == null ) return false ;
      if (getClass()! = obj.getClass()) return false ;
      User other  = (User) obj;
      if (tel  == null ) {
        if (other.tel! = null )
            return false ;
      } else if (! tel.equals(other.tel))
            return false ;
        return true ;
    }
 }
```

4. 窗体设计

窗体设计如图 6 – 2 – 2 所示。

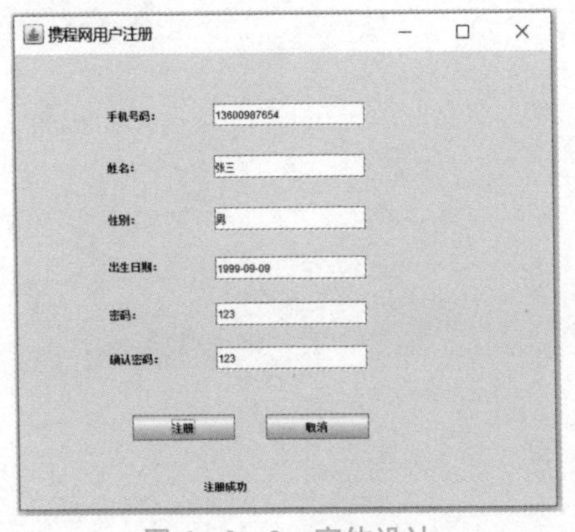


图 6 – 2 – 2　窗体设计

5. 功能实现

代码如下：

```
HashSet<User> userSet = new HashSet<User>();
JButton button = new JButton("注册");
button.addActionListener(new ActionListener() {
      public void actionPerformed(ActionEvent arg0) {
          String tel = textField.getText();
          String name = textField_1.getText();
          String sex = textField_2.getText();
          String birthday = textField_3.getText();
          String pwd = textField_4.getText();
          SimpleDateFormat format = new SimpleDateFormat("yyyy-MM-dd");
          Date birth = null;
          try {
              birth = format.parse(birthday);
          } catch (ParseException e) {
              //TODO Auto-generated catch block
              e.printStackTrace();
          }
          User u = new User(tel,name,sex,birth,pwd);
          userSet.add(u);
          lblNewLabel_4.setText("注册成功");
          System.out.println(userSet);
      }
});
button.setBounds(136, 417, 123, 29);
contentPane.add(button);
```

## 任务工单

| 任务名称 | 图书信息管理系统——读者注册功能 | | | | |
|---|---|---|---|---|---|
| 班级 | | 学号 | | 姓名 | |
| 任务要求 | 读者注册<br>请输入电话号码：<br>请输入姓名：<br>请输入性别：<br>注册　取消<br>（1）定义 HashSet 集合存储读者信息。<br>（2）注册 5 位读者，要求一个手机号码仅能注册一位读者，在控制台输出 5 位读者信息。 | | | | |
| 任务实现 | （提交读者类的设计、Set 集合的定义、界面截图、功能实现相关代码） | | | | |
| 调试过程记录 | | | | | |

# 6.3 Map 接口

## 学习目标

(1) 了解 Map 接口。

(2) 掌握 Map 接口的实现类 HashMap、LinkedHashMap 的应用开发。

(3) 掌握 Properties 集合的应用。

## 相关知识

### 6.3.1 Map 接口

前面所介绍的 Collection 接口是单列集合的父接口，而 Map 接口是双列集合的父接口，它的每个元素由两个部分组成，分别是键与值，键与值成对出现，呈现一种对应的关系，称为映射。在 Map 接口中查找元素，只需要通过键就可以找到对应的值，例如，通过学号就可以找到对应同学的详细信息，通过驾驶证号就可以找到对应车辆的具体信息。

为了实现 Map 接口数据的增、删、改、查，Map 接口提供了一系列方法，如表 6-3-1 所示。

表 6-3-1 Map 接口的常用方法

| 方法 | 功能描述 |
| --- | --- |
| void put(Object key, Object value) | 将指定的值与此映射中的指定键关联（可选操作） |
| Object get(Object key) | 返回指定键所映射的值；如果此映射不包含该键的映射关系，则返回 null |
| boolean containsKey(Object key) | 如果此映射包含指定键的映射关系，则返回 true |
| boolean containsValue(Object value) | 如果此映射将一个或多个键映射到指定值，则返回 true |
| Set keySet() | 返回此映射中包含的键的 Set 视图 |
| Collection <V> values() | 返回此映射中包含的值的 Collection 视图 |
| Set <Map.Entry<K, V>> entrySet() | 返回此映射中包含的映射关系的 Set 视图 |

### 6.3.2 HashMap 集合

HashMap 集合是 Map 接口的一个实现类，是一个存储键值对的集合，每个键值对也叫作 Entry，这些 Entry 分散存储在一个数组中，这个数组也是 HashMap 集合的主干，数组中的元素允许 null 键和 null 值，但由于 key 值必须唯一（不能重复），因此，null 键只有一个。另外，HashMap 集合的元素是无序的，与存入 HashMap 集合的先后顺序无关。

## 一、HashMap 集合的定义

定义 HashMap 集合的语法格式如下：

```
Map[ <key,value> ] hashMap = new HashMap[ <key,value> ]();
```

或

```
HashMap[ <key,value> ] hashMap = new HashMap[ <key,value> ]();
```

（1）<key，value>：可省略。若省略，则 HashMap 集合元素的数据类型不受限制。

（2）key 值必须唯一，若重复存入 key 值相同的元素，则后存入的元素将覆盖前面的元素。

**【例 6-3-1】代码如下：**

```
2  import java.util.*;
3  public class P6_3_1{
4      public static void main(String[] args) {
5          Map hashMap01 = new HashMap();
6          hashMap01.put("1","hello");
7          hashMap01.put("2", 100);
8          HashMap<Integer, String> hashMap02 = new HashMap<Integer, String>();
9          hashMap02.put(1,"张三");
10         hashMap02.put(2,"李四");
11         hashMap02.put(2,"王五");//key 值相同,则"王五"将覆盖"李四"
12         System.out.println(hashMap02);
13         hashMap02.put(2,100); //报错,数据类型不匹配
14         }
15}
```

运行结果如图 6-3-1 所示。

```
{1=张三, 2=王五}
Exception in thread "main" java.lang.Error: Unresolved compilation problem:
	The method put(Integer, String) in the type HashMap<Integer,String> is not applicable for the arguments (int, int)

	at chapter07.Test.main(Test.java:13)
```

图 6-3-1　运行结果

**【例 6-3-2】代码如下：**

```
1  import java.util.*;
2  public class P6_3_2 {
3      public static void main(String[] args) {
4          HashMap<String, Student> stuMap = new HashMap<String, Student>();
5          Student s1 = new Student("001", "林丽娟", "女", 19, "西 001", "外语");
6          Student s2 = new Student("002", "王鹏", "男", 19, null, "土建");
7          Student s3 = new Student("003", "陈萌", "男", 18, "东 003", "信息");
8          stuMap.put(s1.getSid(), s1);//以新生学号作为 key,新生信息为 value
```

```
9           stuMap.put(s2.getSid(), s2);
10          stuMap.put(s3.getSid(), s3);
11          System. out .println(stuMap);
12      }
13  }
```

运行结果如图 6－3－2 所示。

```
{001=Student [sid=001, name=林丽娟, sex=女, age=19, dormId=西001, institute=外语], 002=Student [sid=002, name=王鹏, sex=男,
age=19, dormId=null, institute=土建], 003=Student [sid=003, name=陈萌, sex=男, age=18, dormId=东003, institute=信息]}
```

图 6－3－2 运行结果

【说明】

程序中定义的 HashMap 集合，以新生学号作为键，以新生信息作为值，因此将 <key, value>设置为<String，Student>。

## 二、HashMap 集合键、值的获取

在程序开发中，经常需要获取 Map 接口中的键和值，那么如何实现？有两种方式：第一种方式是先遍历 Map 接口中的所有键，再根据键获取相应的值；第二种方式是先获取 Map 接口中的所有映射关系，然后从映射关系中取出键和值；第三种方式是分别获取 Map 接口中的所有键和值。

### 1. 先遍历所有键，再根据键获取相应的值

1）HashMap 集合获取所有键的方法

Set keySet()：返回此映射中包含键的 Set 视图。

2）HashMap 集合根据键获取值的方法

Object get(Object key)：返回指定键所映射的值；若映射不包含该键的映射关系，则返回 null。

【例 6－3－3】代码如下：

```
1  import java.util. * ;
2  public class P6_3_3{
3      public static void main(String[] args) {
4          HashMap < String, Student > stuMap = new HashMap < String, Student >();
5          Student s1 = new Student("001", "林丽娟", "女", 19, "西 001", "外语");
6          Student s2 = new Student("002", "王鹏", "男", 19, null, "土建");
7          Student s3 = new Student("003", "陈萌", "男", 18, "东 003", "信息");
8          stuMap.put(s1.getSid(), s1);
9          stuMap.put(s2.getSid(), s2);
10         stuMap.put(s3.getSid(), s3);
11         Set keySet = stuMap.keySet(); //获取所有键,返回值类型是 Set 集合
12         System. out .println(keySet);
13         Iterator < String > ite = keySet.iterator();
14         while (ite.hasNext()) {
15             String sid = ite.next();
```

```
16              System.out.println("学号为:" + sid + "的学生信息为:" +
                                        stuMap.get(sid)); //通过键获取相应值
17          }
18      }
19  }
```

运行结果如图6-3-3所示。

```
[001, 002, 003]
学号为: 001的学生信息为: Student [sid=001, name=林丽娟, sex=女, age=19, dormId=西001, institute=外语]
学号为: 002的学生信息为: Student [sid=002, name=王鹏, sex=男, age=19, dormId=null, institute=土建]
学号为: 003的学生信息为: Student [sid=003, name=陈萌, sex=男, age=18, dormId=东003, institute=信息]
```

图6-3-3 运行结果

### 2. 先获取所有映射关系，再从映射关系中取出键和值

1）HashMap集合获取所有映射关系的方法

Set<Map. Entry< K, V> >entrySet()：返回此映射中包含的映射关系的Set视图。

2）Entry获取键的方法

Object getKey()：返回Entry中的键。

3）Entry获取值的方法

Object getValue()：返回Entry中的值。

**【例6-3-4】**代码如下：

```
1  import java.util.*;
2  import java.util.Map.Entry;
3  public class P6_3_4 {
4      public static void main(String[] args) {
5          HashMap<String, Student> stuMap = new HashMap<String, Student>();
6          Student s1 = new Student("001", "林丽娟", "女", 19, "西001", "外语");
7          Student s2 = new Student("002", "王鹏", "男", 19, null, "土建");
8          Student s3 = new Student("003", "陈萌", "男", 18, "东003", "信息");
9          stuMap.put(s1.getSid(), s1);
10         stuMap.put(s2.getSid(), s2);
11         stuMap.put(s3.getSid(), s3);
12         //获取所有键值对
13         Set<Entry<String, Student>> entrySet = stuMap.entrySet();
14         Iterator<Entry<String, Student>> ite = entrySet.iterator();
15         while (ite.hasNext()) {
16             Entry<String, Student> next = ite.next();
17             String key = next.getKey();
18             Student value = next.getValue();
19             System.out.println("学号为:" + key + "的学生信息为:" + value);
20         }
21     }
22 }
```

运行结果如图6-3-4所示。

```
[001, 002, 003]
学号为：001的学生信息为：Student [sid=001, name=林丽娟, sex=女, age=19, dormId=西001, institute=外语]
学号为：002的学生信息为：Student [sid=002, name=王鹏, sex=男, age=19, dormId=null, institute=土建]
学号为：003的学生信息为：Student [sid=003, name=陈萌, sex=男, age=18, dormId=东003, institute=信息]
```

图 6-3-4　运行结果

【说明】

程序第 13 行调用 Map 对象的 entrySet( ) 方法获取存储在 Map 接口中的所有映射，并以 Set 集合的形式返回，Set 集合中存放 Map. Entry 类型的元素（Entry 是 Map 接口的内部接口），每个 Map. Entry 对象代表 Map 接口中的一个键值对。第 16 ~ 18 行获取一个键值对，并调用映射对象的 getKey( ) 方法和 getValue( ) 方法获取键和值。

### 3. 分别获取所有键和所有值

1）HashMap 集合获取所有键的方法

Set keySet( )：返回此映射中包含的键的 Set 视图。

2）HashMap 集合获取所有值的方法

Collection <V> values( )：返回此映射中包含的值的 Collection 视图。

**【6-3-5】**代码如下：

```
1  import java.util.*;
2  import java.util.Map.Entry;
3  public class P6_3_5 {
4      public static void main(String[] args) {
5          HashMap<String, Student> stuMap = new HashMap<String, Student>();
6          Student s1 = new Student("001", "林丽娟", "女", 19, "西 001", "外语");
7          Student s2 = new Student("002", "王鹏", "男", 19, null, "土建");
8          Student s3 = new Student("003", "陈萌", "男", 18, "东 003", "信息");
9          stuMap.put(s1.getSid(), s1);
10         stuMap.put(s2.getSid(), s2);
11         stuMap.put(s3.getSid(), s3);
12         Set keySet = stuMap.keySet();
13         System.out.println("所有的键:" + keySet);
14         Collection<Student> values = stuMap.values();
15         System.out.println("所有的值为:" + values);
16     }
17 }
```

运行结果如图 6-3-5 所示。

```
所有的键：[001, 002, 003]
所有的值为：[Student [sid=001, name=林丽娟, sex=女, age=19, dormId=西001, institute=外语], Student [sid=002, name=王鹏, sex=男, age=19, dormId=null, institute=土建], Student [sid=003, name=陈萌, sex=男, age=18, dormId=东003, institute=信息]]
```

图 6-3-5　运行结果

【说明】

HashMap 集合是无序的，但在实际程序开发中，经常需要一个可以保持插入顺序的 Map 接口。JDK 提供了 LinkedHashMap 集合，它是 HashMap 集合的子类，与 LinkedList 集合一样，使用双向链表来维护内部元素的关系，使 Map 接口元素迭代的顺序与存入的顺序一致。

### 6.3.3　Properties 集合

Map 接口还有一个实现类 HashTable，它与 HashMap 集合十分相似，区别在于 HashTable 集合是线程安全的。HashTable 集合存取元素的速度很低，已经基本被 HashMap 集合所取代，但 HashTable 集合一个子类 Properties，在实际应用中非常重要。Properties 集合的常用方法如表 6-3-2 所示。

表 6-3-2　Properties 集合的常用方法

| 方法 | 功能描述 |
| --- | --- |
| void setProperty(String key, String value) | 调用 HashTable 集合的 put()方法 |
| Object getProperty(String key) | 返回指定键所映射的值 |
| Enumeration PropertyNames() | 返回一个包含所有键的 Enumeration 对象 |
| Set <String> stringPropertyNames() | 返回此属性列表中的键集 |

Properties 集合主要用于存储字符串型的键和值，在实际程序开发中，经常使用 Properties 集合存取应用的配置项。假设有一个数据库连接配置，分别配置数据库驱动、数据库访问地址、访问数据库的用户及密码，具体如下：

```
dataSource.driverClass = com.mysql.jdbc.Driver
dataSource.jdbcUrl = jdbc:mysql://localhost:3306/shop? characterEncoding = UTF-8
dataSource.user = root
dataSource.password = 123456
```

在程序中可以使用 Properties 集合随时访问及修改这些配置项，下面通过【例 6-3-6】来学习 Properties 集合的使用。

**【例 6-3-6】** 代码如下：

```
1  import java.util.*;
2  import java.util.Map.Entry;
3  public class P6_3_6 {
4      public static void main(String[] args) {
5        Properties  p = new Properties();
6        p.setProperty("dataSource.driverClass", "com.mysql.jdbc.Driver");
7        p.setProperty("dataSource.jdbcUrl", "jdbc:mysql://localhost:3306/
                      authorization? characterEncoding = UTF-8");
8        p.setProperty("dataSource.user", "root");
9        p.setProperty("dataSource.password", "123456");
10       Enumeration keys = p.propertyNames();
11       while (keys.hasMoreElements()) {
```

```
12          String key = (String)keys.nextElement();
13          String value = p.getProperty(key);
14          System. out .println(key + " = " + value);
15      }
16    }
17 }
```

运行结果如图 6 -3 -6 所示。

```
dataSource.password=123456
dataSource.user=root
dataSource.driverClass=com.mysql.jdbc.Driver
dataSource.jdbcUrl=jdbc:mysql://localhost:3306/authorization?characterEncoding=UTF-8
```

图 6 -3 -6　运行结果

**案例**

模拟五佳歌手比赛投票，在 10 位参赛选手中选择 5 位喜欢的歌手，为他们投票。系统允许每个用户只提交一次投票结果。投票结束后，系统自动统计投票结果。系统如何存储所有用户的投票结果及所有歌手的得票情况？

【分析】

(1) 所有用户的投票结果采用 Map 接口存储，用户名为键，用户的投票结果为值。每个用户选择的 5 位歌手采用 Set 接口存储，这里不能使用 List 接口，因为 List 接口允许重复值，这样会造成一个用户为一个歌手投两次票。

(2) 记录所有歌手的得票结果，采用 Map 接口存储，歌手姓名为键，歌手信息（包含得票结果）为值。

**1. 类的定义**

代码如下：

```
public class Singer {
      private String name ;
      private String song;
      private int poll;//得票数
      //省略 get()方法、set()方法、构造方法、toString()方法
}
```

**2. 集合的设计**

(1) voteMap 集合：用于存放所有用户的投票信息。

(2) singerMap 集合：用于存放歌手信息及得票情况。

**3. 窗体设计**

窗体设计如图 6 -3 -7 所示。

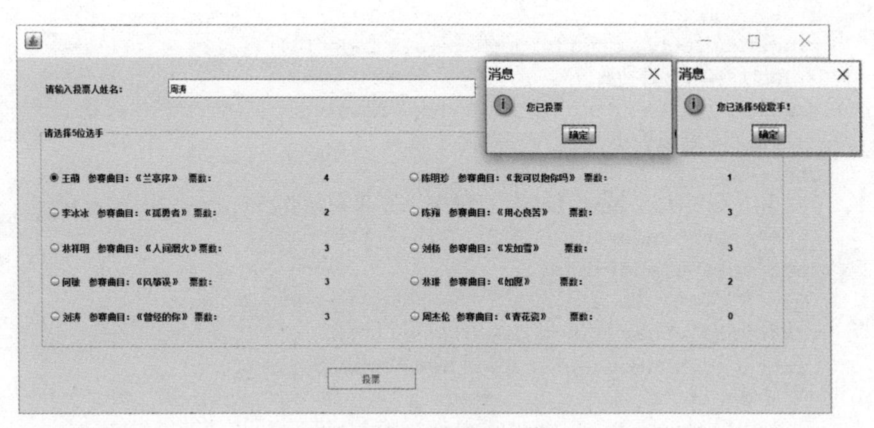


图6-3-7　窗体设计

### 4. 功能实现

代码如下:

```
public class VoteFrame extends JFrame {
    private JPanel contentPane;
    private JTextField textField;
    int cnt;
    Set<String> selectSet;
    JButton button;
    ……(略)
    public VoteFrame() {
        ……(略)
        //歌手信息及得票情况,在初始状态下得票为0
        Map<String,Singer> singerMap = new HashMap<String,Singer>();
        Singer singer1 = new Singer("王萌","《兰亭序》",0);
        Singer singer2 = new Singer("李冰冰","《孤勇者》",0);
        Singer singer3 = new Singer("林祥明","《人间烟火》",0);
        Singer singer4 = new Singer("何敏","《风筝误》",0);
        Singer singer5 = new Singer("刘涛","《曾经的你》",0);
        Singer singer6 = new Singer("陈明珍","《我可以抱你吗》",0);
        Singer singer7 = new Singer("陈翔","《用心良苦》",0);
        Singer singer8 = new Singer("刘杨","《发如雪》",0);
        Singer singer9 = new Singer("林琳","《如愿》",0);
        Singer singer10 = new Singer("周杰伦","《青花瓷》",0);
        //歌手信息及得票情况存储在Map接口中
        singerMap.put(singer1.getName(), singer1);
        singerMap.put(singer2.getName(), singer2);
        singerMap.put(singer3.getName(), singer3);
        singerMap.put(singer4.getName(), singer4);
        singerMap.put(singer5.getName(), singer5);
        singerMap.put(singer6.getName(), singer6);
        singerMap.put(singer7.getName(), singer7);
        singerMap.put(singer8.getName(), singer8);
        singerMap.put(singer9.getName(), singer9);
```

```
singerMap.put(singer10.getName(), singer10);
//存储所有用户的投票情况
Map<String,Set<String>> voteMap = new HashMap<String, Set<String>>();
//每位用户选择歌手数统计(不超过5位)
cnt = 0;
JLabel label = new JLabel("请输入投票人姓名:");
label.setBounds(40, 33, 182, 21);
contentPane.add(label);

textField = new JTextField();
textField.addFocusListener(new FocusAdapter() {
    @Override
    public void focusLost(FocusEvent arg0) {
        String username = textField.getText();
        Set<String> users = voteMap.keySet();//获取所有已投票用户
        System.out.println(users);
        if (users.contains(username)) {//判断当前用户是否在已投票列表中
          JOptionPane.showMessageDialog(null, "您已投票");
          textField.setText("");
      }else
          button.setEnabled(true);
    }
});
textField.setBounds(214, 30, 432, 27);
contentPane.add(textField);
textField.setColumns(10);
……(略)

JRadioButton rb = new JRadioButton("王萌 参赛曲目:《兰亭序》票数:");
  rb.addActionListener(new ActionListener() {
  public void actionPerformed(ActionEvent arg0) {
    if (rb.isSelected()) //用户选择5位歌手后,不允许继续选择
      if (cnt! =5){
        cnt ++;
    }else {
        JOptionPane.showMessageDialog(null,"您已选5位歌手!");
        rb.setSelected(false);
        }
    }
});
rb.setBounds(11, 56, 375, 29);
panel.add(rb);

……(略)
JRadioButton rb9 = new JRadioButton("周杰伦 参赛曲目:《青花瓷》票数:");
rb9.addActionListener(new ActionListener() {
   public void actionPerformed(ActionEvent arg0) {
     if (rb9.isSelected()) {
       if (cnt! =5){
```

```
            cnt ++;
        }else {
            JOptionPane.showMessageDialog(null,"您已选5位歌手!");
            Rb9.setSelected(false);
        }
      }
    }
});
rb9.setBounds(518, 247, 444, 29);
panel.add(rb9);
//歌手选票总票数展示标签
JLabel label_1 = new JLabel("0");
label_1.setBounds(402, 60, 51, 21);
panel.add(label_1);
JLabel label_1_1 = new JLabel("0");
label_1_1.setBounds(402, 108, 51, 21);
panel.add(label_1_1);
……(略)
JLabel label_1_9 = new JLabel("0");
label_1_9.setBounds(970, 251, 51, 21);
panel.add(label_1_9);

button = new JButton("投票");
button.addActionListener(new ActionListener() {
    public void actionPerformed(ActionEvent arg0) {
      String username = textField.getText();
      //存储用户选择的歌手集合
      selectSet = new HashSet<String>();
      if (rb.isSelected()) {
      selectSet.add("王萌");//将被选歌手姓名添加到selectSet集合中
      int n = singerMap.get("王萌").getPoll();//获取歌手得票数
        singerMap.get("王萌").setPoll(n+1);//歌手得票数增1
        //更改窗体中歌手总票数
        label_1.setText(singerMap.get("王萌").getPoll() + "");
    }
    if (rb1.isSelected()) {
        selectSet.add("李冰冰");
        int n = singerMap.get("李冰冰").getPoll();
        singerMap.get("李冰冰").setPoll(n+1);
        label_1_1.setText(singerMap.get("李冰冰").getPoll() + "");
    }
    ……(略)
    if (rb9.isSelected()) {
        selectSet.add("周杰伦");
        int n = singerMap.get("周杰伦").getPoll();
        singerMap.get("周杰伦").setPoll(n+1);
        label_1_9.setText(singerMap.get("周杰伦").getPoll() + "");
  }
  //将当前用户选择的歌手集合存入总投票集合voteMap中
```

```
            voteMap.put(username,selectSet); System. out .println(singerMap);
            System. out .println(voteMap);
            //操作界面清理
            button.setEnabled(false );
            textField.setText("");
            radioButton.setSelected(false );
            radioButton_1.setSelected(false );
            ……(略)
            radioButton_9.setSelected(false );
            cnt =0;
        }
    });
    button.setBounds(436, 428, 123, 29);
    contentPane.add(button);
  }
}
```

### 任务工单

<table>
<tr><td>任务名称</td><td colspan="5">图书信息管理系统——读者注册功能</td></tr>
<tr><td>班级</td><td></td><td>学号</td><td></td><td>姓名</td><td></td></tr>
<tr><td>任务要求</td><td colspan="5">读者注册<br>请输入电话号码：<br>请输入姓名：<br>请输入性别：<br>注册　取消<br>（1）定义 HashMap 集合存储读者信息。<br>（2）注册 5 位读者，要求一个手机号码仅能注册一位读者，在控制台输出 5 位读者的信息。</td></tr>
<tr><td>任务实现</td><td colspan="5">[提交实现任务（1）、（2）功能的代码]</td></tr>
<tr><td>调试过程记录</td><td colspan="5"></td></tr>
</table>

## 6.4 集合与 JTable

### 学习目标

（1）掌握 ArrayList、LinkedList、HashSet、TreeSet 、HashMap 等集合的应用。
（2）掌握 JTable 的构造方法及常用方法。
（3）掌握用 JTable 展示集合数据的方法。

### 相关知识

JTable 是 GUI 程序中常用的组件，使用它可以创建一个由多行、多列组成的二维表格。一个表格由两部分组成，分别为表体及表头，表体用于显示表格数据，它既可以是二维数组，也可以是 Vector 集合（Vector 集合包含其形成的二维数据）；表头用于显示表格列标题，它既可以是一维数组，也可以是 Vector 集合。

### 一、创建表格

Java 为创建表格提供了 7 个构造方法，如表 6 - 4 - 1 所示，根据不同的参数所创建的表格不同。

表 6 - 4 - 1 JTable 的构造方法

| 方法 | 功能描述 |
| --- | --- |
| JTable() | 使用系统默认的 Model 创建一个 JTable |
| JTable(int numRows, int numColumns) | 使用 DefaultTableModel 创建一个具有 numRows 行、numColumns 列的空表格 |
| JTable(Object[][] rowData, Object[] columnNames) | 创建一个以二维数组 rowData 为数据、以二维数组 columnNames 为标题的表格 |
| JTable(TableModel dm) | 创建一个 JTable，具有默认的字段模式以及选择模式，并设置数据模式 |
| JTable(TableModel dm, TableColumnModel cm) | 创建一个 Jtable，已设置数据模式与字段模式，并具有默认的选择模式 |
| JTable(TableModel dm, TableColumnModel cm, ListSelectionModel sm) | 创建一个 JTable，已设置数据模式、字段模式与选择模式 |
| JTable(Vector rowData, Vector columnNames) | 创建一个以 Vector 集合为数据源的表格，可显示列标题 |

#### 1. JTable（Object[][] rowData，Object[] columnNames）构造方法

在表格数据相对固定的情况下，使用数组作为表格参数。JTable 的表体为一个二维数

组，表头为一个一维数组。

**【例6-4-1】** 代码如下：

```
1  public class ArrayTable extends JFrame {
2      private JPanel contentPane;
3      private JTable table;
4      public static void main(String[] args) {
5          EventQueue.invokeLater(new Runnable() {
6              public void run() {
7                  try {
8                      ArrayTable frame = new ArrayTable();
9                      frame.setVisible(true);
10                 } catch (Exception e) {
11                     e.printStackTrace();
12                 }
13             }
14         });
15 }
16 public ArrayTable() {
17     setDefaultCloseOperation(JFrame.EXIT_ON_CLOSE);
18     setBounds(100, 100, 983, 325);
19     contentPane = new JPanel();
20     contentPane.setBorder(new EmptyBorder(5, 5, 5, 5));
21     setContentPane(contentPane);
22     contentPane.setLayout(null);
23     JScrollPane scrollPane = new JScrollPane();
24     scrollPane.setBounds(15, 15, 931, 239);
25     contentPane.add(scrollPane);
26     //定义表体(二维数组)
27     Object[][] data = {
28         new Object[] {"001", "林丽娟", "女", 19, "西001", "外语"},
29         new Object[] {"002", "王鹏", "男", 19, null, "土建"},
30         new Object[] {"003", "陈萌", "男", 18, "东003", "信息"}
31     };
32     //定义表头(一维数组)
33     Object[] title = {"学号","姓名","性别","年龄","宿舍","所在学院"};
34     //创建表格
35     table = new JTable(data, title);
36     scrollPane.setViewportView(table);
37   }
38 }
```

运行结果如图6-4-1所示。

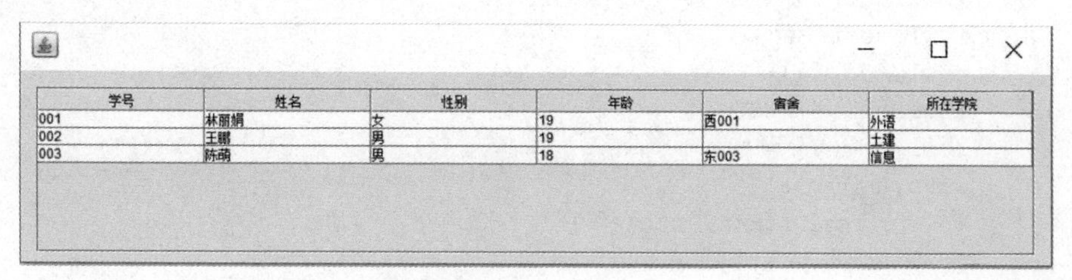

| 学号 | 姓名 | 性别 | 年龄 | 宿舍 | 所在学院 |
|---|---|---|---|---|---|
| 001 | 林丽娟 | 女 | 19 | 西001 | 外语 |
| 002 | 王鹏 | 男 | 19 | | 土建 |
| 003 | 陈萌 | 男 | 18 | 东003 | 信息 |

图6-4-1　运行结果

【说明】

从运行结果看，表格共 3 行、6 列，程序中定义了一个二维数组 data 作为表体，一个一维数组 title 作为表头，调用 JTable（Object[ ][ ] rowData，Object[ ]columnNames）构造方法创建表格。

**2. JTable（Vector rowData，Vector columnNames）构造方法**

由于集合更易于维护，所以表格需要频繁更新时采用集合作为表格参数。JTable 的表体及表头均为一个 Vector 集合。

**【例 6-4-2】** 代码如下：

```
1  public class VectorTable extends JFrame {
2      private JPanel contentPane;
3      private JTable table;
4      public static void main(String[] args) {
5          EventQueue.invokeLater(new Runnable() {
6              public void run() {
7                  try {
8                      VectorTable frame = new VectorTable();
9                      frame.setVisible(true);
10                 } catch (Exception e) {
11                     e.printStackTrace();
12                 }
13             }
14         });
15     }
16     public VectorTable() {
17         setDefaultCloseOperation(JFrame.EXIT_ON_CLOSE);
18         setBounds(100, 100, 849, 352);
19         contentPane = new JPanel();
20         contentPane.setBorder(new EmptyBorder(5, 5, 5, 5));
21         setContentPane(contentPane);
22         contentPane.setLayout(null);
23         JScrollPane scrollPane = new JScrollPane();
24         scrollPane.setBounds(15, 25, 797, 256);
25         contentPane.add(scrollPane);
26         //定义表体(集合)
27         Vector data = new Vector();
28         Vector row1 = new Vector();
29         row1.add("001");
30         row1.add("林丽娟");
31         row1.add("女");
32         row1.add(19);
33         row1.add("西 001");
34         row1.add("外语");
35         data.add(row1);
36         Vector row2 = new Vector();
37         row2.add("001");
38         row2.add("王鹏");
```

```
39          row2.add("男");
40          row2.add("19");
41          row2.add(null);
42          row2.add("土建");
43          data.add(row2);
44          //定义表头(集合)
45          Vector<String> title = new Vector<String>();
46          title.add("学号");
47          title.add("姓名");
48          title.add("性别");
49          title.add("年龄");
50        title.add("宿舍");
51          title.add("所在学院");
52          //创建表格
53          table = new JTable(data,title);
54          scrollPane.setViewportView(table);
55      }
56  }
```

运行结果如图6－4－2所示。

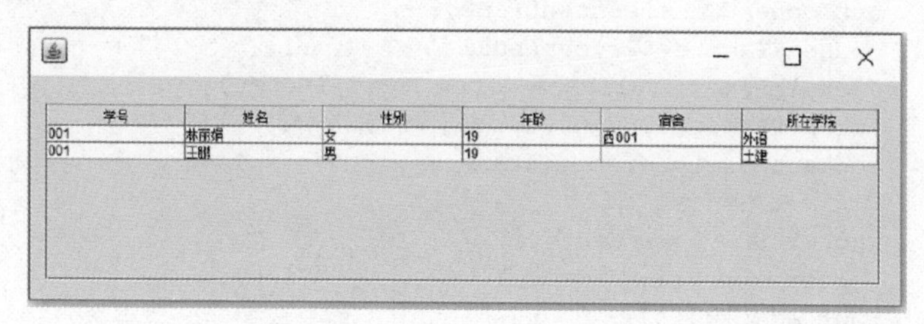

学号
姓名
性别
年龄
宿舍
所在学院
001
女
19
西001
001
男
19
土建


图6－4－2　运行结果

【说明】

从运行结果看，表格共2行、6列，程序第27行定义了一个Vector集合data作为表体，第28、36行定义了两个Vector集合row1、row2用于存储表格的两行数据，第35、43行将两行数据添加到集合data中，此时集合data升级为二维。第45行定义表头集合title，调用JTable(Vector rowData，Vector columnNames）构造方法创建表格。

### 3. JTable(TableModel dm）构造方法

TableModel是一个接口，其中定义了存取表格单元格内容、计算表格列数等存取表格的抽象方法，实现了这些抽象方法就可以确定一个表格。Java提供了两个实现TableModel接口的类。

1）AbstractTableMode抽象类

AbstractTableMode抽象类实现了TableModel接口中的大部分方法——除了getRowCount()、getColumnCount()、getValueAt()3个方法。使用它创建表格，只要实现这3个方法即可，为了获取列标题，还必须实现getColumnName()方法。

【例6－4－3】代码如下：

```
1  public class TableModelTable extends JFrame {
2      private JPanel contentPane;
3      private JTable table;
4      public static void main(String[] args) {
5          EventQueue.invokeLater(new Runnable() {
6            public void run() {
7                try {
8                    VectorTable frame = new VectorTable();
9                    frame.setVisible(true);
10               } catch (Exception e) {
11                   e.printStackTrace();
12               }
13           }
14       });
15     }
16     public VectorTable() {
17         setDefaultCloseOperation(JFrame.EXIT_ON_CLOSE);
18         setBounds(100, 100, 849, 352);
19         contentPane = new JPanel();
20         contentPane.setBorder(new EmptyBorder(5, 5, 5, 5));
21         setContentPane(contentPane);
22         contentPane.setLayout(null);
23         JScrollPane scrollPane = new JScrollPane();
24         scrollPane.setBounds(15, 25, 797, 256);
25         contentPane.add(scrollPane);
26         //定义表体(集合)
27         Vector data = new Vector();
28         Vector row1 = new Vector();
29         row1.add("001");
30         row1.add("林丽娟");
31         row1.add("女");
32         row1.add(19);
33         row1.add("西001");
34         row1.add("外语");
35         data.add(row1);
36         Vector row2 = new Vector();
37         row2.add("001");
38         row2.add("王鹏");
39         row2.add("男");
40         row2.add("19");
41         row2.add(null);
42         row2.add("土建");
43         data.add(row2);
44         //定义表头(集合)
45         Vector<String> title = new Vector<String>();
46         title.add("学号");
47         title.add("姓名");
48         title.add("性别");
49         title.add("年龄");
```

```
50          title.add("宿舍");
51          title.add("所在学院");
52
53          AbstractTableModel model = new AbstractTableModel() {
54              @ Override//获取指定单元格的值
55              public Object getValueAt(int rowIndex, int columnIndex) {
56                  //TODO Auto-generated method stub
57                  return ((Vector)data.get(rowIndex)).get(columnIndex);
58              }
59              @ Override//获取列数
60              public int getColumnCount() {
61                  //TODO Auto-generated method stub
62                  return title.size();
63              }
64              @ Override//获取行数
65              public int getRowCount() {
66                  //TODO Auto-generated method stub
67                  return data.size();
68              }
69              //获取列标题
70              public String getColumnName(int column){
71                  return (String)title.get(column);
72              }
73          };
74          table = new JTable();
75          scrollPane.setViewportView(table);
76      }
77  }
```

运行结果如图6-4-3所示。

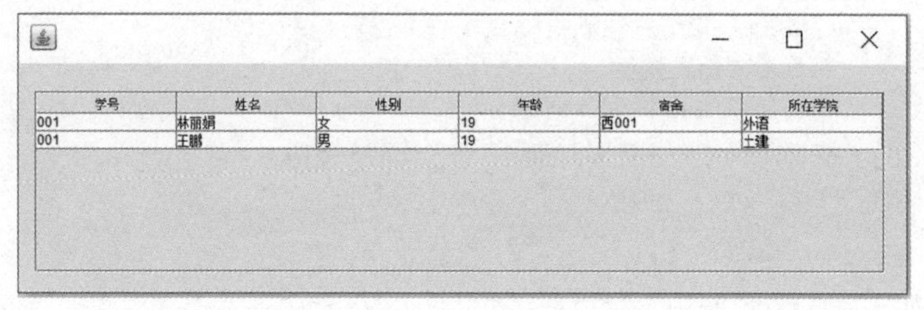

| 学号 | 姓名 | 性别 | 年龄 | 宿舍 | 所在学院 |
| --- | --- | --- | --- | --- | --- |
| 001 | 林丽娟 | 女 | 19 | 西001 | 外语 |
| 001 | 王鹏 | 男 | 19 |  | 土建 |

图6-4-3 运行结果

2）DefaultTableMode 实现类

DefaultTableMode 实现类继承 AbstractTableMode 抽象类，是默认的表格模式。

**【例6-4-4】** 代码如下：

```
1  public class TableModelTable extends JFrame {
2      private JPanel contentPane;
3      private JTable table;
4      public static void main(String[] args) {
```

```
5           EventQueue.invokeLater(new Runnable() {
6               public void run() {
7                   try {
8                       ArrayTable frame = new ArrayTable();
9                       frame.setVisible(true);
10                  } catch (Exception e) {
11                      e.printStackTrace();
12                  }
13              }
14          });
15      }
16      public TableModelTable () {
17          setDefaultCloseOperation(JFrame.EXIT_ON_CLOSE);
18          setBounds(100, 100, 983, 325);
19          contentPane = new JPanel();
20          contentPane.setBorder(new EmptyBorder(5, 5, 5, 5));
21          setContentPane(contentPane);
22          contentPane.setLayout(null);
23          JScrollPane scrollPane = new JScrollPane();
24          scrollPane.setBounds(15, 15, 931, 239);
25          contentPane.add(scrollPane);
26          //定义表体(二维数组)
27          Object[][] data = {
28              new Object[] {"001", "林丽娟", "女", 19, "西 001", "外语"},
29              new Object[] {"002", "王鹏", "男", 19, null, "土建"},
30              new Object[] {"003", "陈萌", "男", 18, "东 003", "信息"}
31          };
32            //定义表头(一维数组)
33            Object[] title = {"学号","姓名","性别","年龄","宿舍","所在学院"};
34            //定义一个 DefaultTableModel 模型,为模型设置填充数据
35            DefaultTableModel m = new DefaultTableModel();
36            m.setDataVector(data1, title1);
37            //为表格设置模型
38            table = new JTable(m);
39            scrollPane.setViewportView(table);
40    }
41  }
```

运行结果如图 6-4-4 所示。

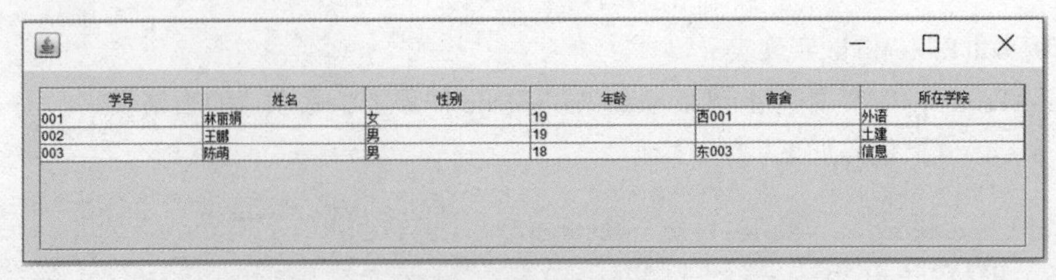


图 6-4-4　运行结果

## 二、表格常用方法

在程序开发中，需要对表格进行数据存取，Java 提供了表 6－4－2 所示的方法来完成表格操作。

表 6－4－2　JTable 的常用方法

| 方法 | 功能描述 |
| --- | --- |
| TableModel getModel( ) | 获取 JTable 数据模式 |
| void setModel( TableModel dataModel) | 设置 JTable 数据模式为 dataModel |
| int getColumnCount( ) | 获取 JTable 列数 |
| int getRowCount( ) | 获取 JTable 行数 |
| Object getValueAt( int row, int column) | 获取 JTable 指定单元格的数据 |
| void setValueAt( Object value, int row, int column) | 设置 JTable 指定单元格值为 value |
| int getSelectedColumn( ) | 获取用户选定列的列号 |
| int getSelectedRow( ) | 获取用户选定行的行号 |
| int getSelectedColumnCount( ) | 获取用户选定列的列数 |
| int getSelectedRowCount( ) | 获取用户选定行的行数 |

在程序开发过程中，表格中的数据会实时更新，Java 为 JTable 提供了一个更改表格数据的方法 setModel( TableModel model)，它通过重新设置表格模型来更新表格。

**【例 6－4－5】**代码如下：

```
1    public TableModelTable () {
2        setDefaultCloseOperation(JFrame.EXIT_ON_CLOSE );
3        setBounds(100, 100, 983, 325);
4        contentPane = new JPanel();
5        contentPane.setBorder(new EmptyBorder(5, 5, 5, 5));
6        setContentPane(contentPane);
7        contentPane.setLayout(null );
8        JScrollPane scrollPane = new JScrollPane();
9        scrollPane.setBounds(15, 15, 931, 239);
10        contentPane.add(scrollPane);
11        //定义表体(二维数组)
12        Object[][] data = new Object[5][6];
13        data[0] =new Object[] {"001", "林丽娟", "女", 19, "西 001", "外语"},
14        data[1] =new Object[] {"002", "王鹏", "男", 19, null , "土建"},
15        data[2] =new Object[] {"003", "陈萌", "男", 18, "东 003", "信息"}
16        //定义表头(一维数组)
17        Object[] title = {"学号","姓名","性别","年龄","宿舍","所在学院"};
18        table = new JTable(data,title);
```

```
19        scrollPane.setViewportView(table);
20
21        JButton button_1 = new JButton("添加一行数据");
22        button_1.addActionListener(new ActionListener() {
23            public void actionPerformed(ActionEvent arg0) {
24                //修改二维数组
25                data[3] = new Object[] {"004", "胡洋", "男", 18, "东 003", "信息"};
26                //刷新表格数据
27                DefaultTableModel model = new DefaultTableModel();
28                model.setDataVector(data, title);
29                table.setModel(model);
30            }
31          });
32          button_1.setBounds(42, 200, 128, 23);
33          contentPane.add(button_1);
34
35          JButton button = new JButton("打印当前选定行");
36          button.addActionListener(new ActionListener() {
37              public void actionPerformed(ActionEvent e) {
38                  int row = table.getSelectedRow();//获取选定行号
39                  int column = table.getSelectedColumn();//获取选定列号
40                  //获取指定单元格数据
41                  lblNewLabel.setText("当前选定单元格:" +
                              table.getValueAt(row,column).toString());
42              }
43          });
44          button.setBounds(200, 200, 156, 23);
45          contentPane.add(button);
46          lblNewLabel = new JLabel("");
47          lblNewLabel.setBounds(52, 250, 304, 15);
48          contentPane.add(lblNewLabel);
49      }
50  }
```

运行结果如图 6-4-5 所示。

| 学号 | 姓名 | 性别 | 年龄 | 宿舍 | 所在学院 |
|---|---|---|---|---|---|
| 001 | 林丽娟 | 女 | 19 | 西001 | 外语 |
| 002 | 王鹏 | 男 | 19 |  | 土建 |
| 003 | 陈萌 | 男 | 18 | 东003 | 信息 |
| 004 | 胡洋 | 男 | 18 | 东003 | 信息 |

添加一行数据　　打印当前选定单元格数据

当前选定单元格：东003

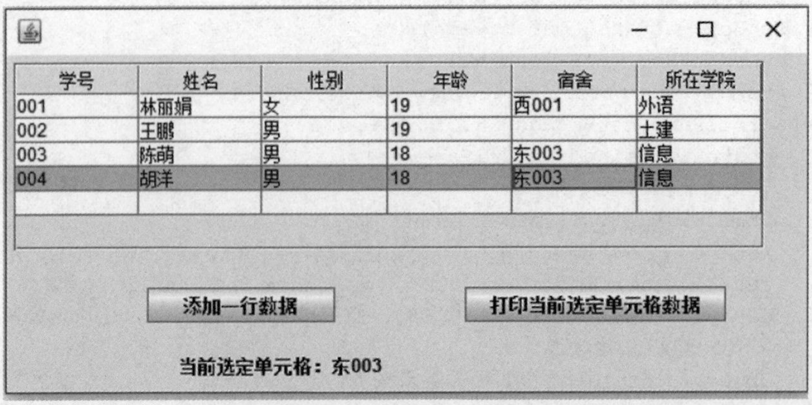

图 6-4-5　运行结果

【说明】

单击“添加一行数据”按钮，则在二维数组 data 中添加 data[3]元素，但表格数据的刷新必须调用 setModel(TableModel dataModel) 方法来实现，因此在程序第 27 行定义了一个数据模型，第 28 行为数据模型设置表体数组 data 和表头数组 title，第 29 行重新定义表格模型。单击“打印当前选定单元格数据”按钮，则通过 getSelectedRow()、getSelectedColumn()方法获取当前选定行号及列号，并调用 getValueAt(row, column) 方法获取指定单元格数据。

**案例**

设计小区物业管理系统水电费管理模块（图 6-4-6），要求实现水电费录入、水电费缴费、水电费查询、水电费信息更新及删除、未缴费情况统计、催缴等功能。在未学习数据库之前，大量的水电费信息存储在集合中，系统交互界面采用表格的形式完成交互。

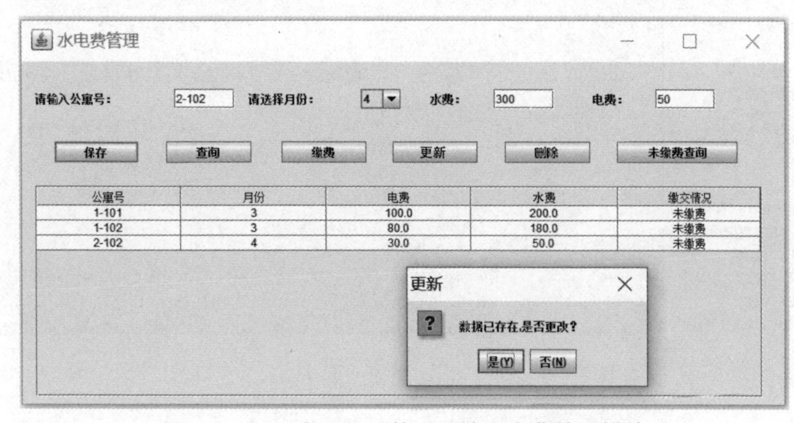


图 6-4-6　小区物业管理系统水电费管理模块

【分析】

(1) 每个公寓每个月都会产生相应的水电费，因此定义封装类 HouseFee，包含公寓号、月份、当月水费、当月电费、缴费状态。

(2) 小区所有的水电费都保存在集合中，集合的增、删、改操作较数组来说更简单，集合采用 List 接口、Set 接口、Map 接口均可。

(3) 使用 JTable(Vector rowData, Vector columnNames) 创建表格，使用 void setModel (TableModel dataModel) 方法实时更新表格数据。

**1. 类的定义**

代码如下：

```
public class HouseFee {
      private String houseid;
      private String month;
      private double elecFee;
      private double waterFee;
      private String status;
      //省略 get()方法、set()方法、构造方法、toString()方法
}
```

### 2. 集合的设计

（1）feeList 集合（ArrayList）：用于存放所有用户的水电费信息。

（2）unpaidList 集合（ArrayList）：用于存放所有未缴交的水电费信息。

（3）resultList 集合（ArrayList）：用于存放集合中满足条件的结果。

（4）data 集合（Vector）：表格内容（二维）。

（5）columnNames 集合（Vector）：表格表头（一维）。

### 3. 窗体设计

窗体设计如图 6－4－7 所示。

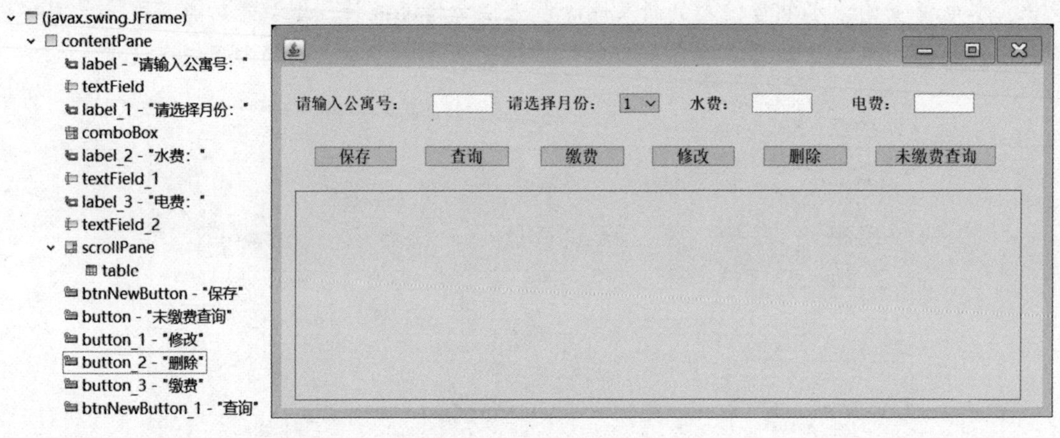


图 6－4－7　窗体设计

### 4. 工具类

代码如下：

```
public class TableUtil {
    public static Vector getDate(List<HouseFee> list) {
        Vector data = new Vector();
        for (HouseFee HouseFee : list) {
            Vector row = new Vector();
            row.add(HouseFee.getHouseid());
            row.add(HouseFee.getMonth());
            row.add(HouseFee.getElecFee());
            row.add(HouseFee.getWaterFee());
            row.add(HouseFee.getStatus());
            data.add(row);
        }
    return data;
  }
}
```

### 5. 功能实现

（1）保存功能：录入公寓号、月份、水费、电费等信息，保存至 feeList 集合中。当出现重复录入同一个公寓同一月份的水电费信息的情况时，要求提示是否保存，若保存则新数

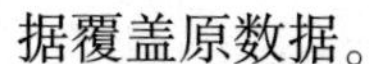

据覆盖原数据。

（2）查询功能：按照公寓号或月份信息在 feeList 集合中查找相关水电费信息。

（3）缴费功能：选中表格中的某行数据，单击“缴费”按钮，则更改缴费状态为“已缴费”。

（4）更新功能：选中表格中的某行数据，单击“更新”按钮，则该行数据显示在窗体顶端的文本框中，修改后单击“保存”按钮，则完成更新。

（5）删除功能：选中表格中的某行数据，单击“删除”按钮，则 feeList 集合删除该条数据，同时表格中删除该条数据。

（6）未缴费查询功能：查询所有未缴费的水电费数据，在表格中显示出来。

代码如下：

```
public class FeeFrame extends JFrame {
      private JPanel contentPane;
      private JTextField textField, textField_1,textField_2;
      private JTable table;
      ……(略)
      public FeeFrame() {
          setcolumnNames("水电费管理");
          ……(略)
          table = new JTable();
          scrollPane.setViewportView(table);
          //所有水电费集合
          List<HouseFee> feeList = new ArrayList<HouseFee>();
          Vector data = new Vector();//表格数据
          Vector columnNames = new Vector();//表头
          columnNames.add("公寓号"); columnNames.add("月份");
          columnNames.add("电费"); columnNames.add("水费");
          columnNames.add("缴交情况");
          DefaultTableModel model = new DefaultTableModel();
          //设置表格数据居中对齐
          DefaultTableCellRenderer tcr = new DefaultTableCellRenderer();
          tcr.setHorizontalAlignment(SwingConstants. CENTER);
          table.setDefaultRenderer(Object. class, tcr);
          JButton btnNewButton = new JButton("保存");//录入水电费信息功能
          btnNewButton.addActionListener(new ActionListener() {
              public void actionPerformed(ActionEvent arg0) {
              String hid = textField.getText();
              String month = comboBox.getSelectedItem().toString();
              double elec = Double.parseDouble(textField_1.getText());
              double water = Double.parseDouble(textField_2.getText());
              HouseFee hf = new HouseFee(hid,month,elec,water,"未缴费");
              Iterator<HouseFee> ite = feeList.iterator();
              while (ite.hasNext()) {
                  HouseFee houseFee = ite.next();
                  //判断数据是否已经存在
                  if (hid.equals(houseFee.getHouseid()) &&
```

```
                              month.equals(houseFee.getMonth())) {
                  //弹出提示框
                  int index = JOptionPane.showConfirmDialog(null, "数据已
                  存在,是否更改?","更新",JOptionPane.YES_NO_OPTION);
                  if (index == 0) {
                     houseFee.setElecFee(elec);
                     houseFee.setWaterFee(water);
                     //调用工具类,将 ArrayList 集合转换为表体 Vector 集合
                     Vector data = TableUtil.getDate(feeList);
                     //修改 TalbeModel 的数据,刷新表格
                     model.setDataVector(data, columnNames);
                     table.setModel(model);
                  }
                  return;
               }
            }
            feeList.add(hf); //水电费集合中增加新数据
            Vector row = new Vector();
            row.add(hid); row.add(month);
            row.add(elec); row.add(water); row.add("未缴费");
            data.add(row);
            model.setDataVector(data, columnNames);
            table.setModel(model);
         }
      });
      btnNewButton.setBounds(36, 85, 93, 23);
      contentPane.add(btnNewButton);
      List<HouseFee> unpaidList = new ArrayList<HouseFee>(); //定义未缴费集合
      JButton button  = new JButton("未缴费查询"); //未缴费查询功能
      button.addActionListener(new ActionListener() {
        public void actionPerformed(ActionEvent arg0) {
          for (HouseFee houseFee : feeList) {
            if (houseFee.getStatus().equals("未缴费"))
               unpaidList.add(houseFee);
          }
          Vector unpaidData = TableUtil.getDate(unpaidList);
          model.setDataVector(unpaidData, columnNames);
          table.setModel(model);
        }
      });
   button.setBounds(652, 85, 134, 23);
   contentPane.add(button);
   JButton button_1  = new JButton("更新"); //更新水电费信息功能
   button_1.addActionListener(new ActionListener() {
      public void actionPerformed(ActionEvent e) {
        //获取表格选定行第 1 个单元格的值
        String hid = table.getValueAt(table.getSelectedRow(), 0).toString();
        String month = table.getValueAt(table.getSelectedRow(), 1).toString();
        String elec = table.getValueAt(table.getSelectedRow(), 2).toString();
```

```
        String water = table.getValueAt(table.getSelectedRow(), 3).toString();
        String status = table.getValueAt(table.getSelectedRow(),4).toString();
        textField.setText(hid);
        textField_1.setText(elec);
        textField_2.setText(water);
        comboBox.setSelectedItem(month);
      }
    });
    button_1.setBounds(407, 85, 93, 23);
    contentPane.add(button_1);
    JButton button_2 = new JButton("删除");//删除水电费信息功能
    button_2.addActionListener(new ActionListener() {
      ……(略)

    });
    button_2.setBounds(530, 85, 93, 23);
    contentPane.add(button_2);
    JButton button_3 = new JButton("缴费");//水电费缴交功能
    button_3.addActionListener(new ActionListener() {
        ……(略)
    });
    button_3.setBounds(285, 85, 93, 23);
    contentPane.add(button_3);
    List<HouseFee> resultList = new ArrayList<HouseFee>();//定义查询结果集合
    JButton btnNewButton_1 = new JButton("查询");
    btnNewButton_1.addActionListener(new ActionListener() {
        ……(略)
    });
    btnNewButton_1.setBounds(159, 85, 93, 23);
    contentPane.add(btnNewButton_1);
}}
```

## 任务工单

<table>
<tr><td>任务名称</td><td colspan="5">图书信息管理系统——读者信息模糊查询</td></tr>
<tr><td>班级</td><td></td><td>学号</td><td></td><td>姓名</td><td></td></tr>
<tr><td>任务要求</td><td colspan="5">读者信息模糊查询<br>关键字： 王　模糊查询<br>tel | name | sex<br>15960980123 | 王影 | 女<br>13598767877 | 王培峰 | 男<br>13600989876 | 王强 | 男<br><br>（1）定义 HashMap 集合存储读者信息，代码如下：<br><br>Map < String,Reader > readers = new HashMap < String,Reader >();<br>readers.put("13600989876",new Reader("13600989876","王强","男"));<br>readers.put("13598767877",new Reader("13598767877","王培峰","男"));<br>readers.put("15960980123",new Reader("15960980123","王影","女"));<br>readers.put("15256789876",new Reader("15256789876","周红梅","女"));<br>readers.put("13798098088",new Reader("13798098088","林琳","女"));<br>readers.put("13850987123",new Reader("13850987123","林星月","男"));<br><br>（2）实现根据姓名模糊查询读者信息的功能。</td></tr>
<tr><td>任务实现</td><td colspan="5"></td></tr>
<tr><td>调试过程记录</td><td colspan="5"></td></tr>
</table>

## 练习题

### 1. 选择题

（1）下列集合中不属于 Collection 接口的子类的是（　　）。

A. ArrayList　　B. LinkedList　　C. TreeSet　　D. Properties

（2）下列关于 List 接口的描述中错误的是（　　）。

A. List 接口是有索引的

B. List 接口可以存储重复的元素

C. List 接口存和取的顺序一致

D. List 接口没有索引

（3）执行下面的程序，其运行结果是（　　）。

```
public class Example {
    public static void main(String[] args) {
        ArrayList list = new ArrayList();  //创建 ArrayList 集合
        list.add("Tom");    //向该集合中添加字符串
        list.add("Jerry");
        Iterator it = list.iterator();  //获取迭代器
        it.next();
        it.next();
System.out.println(it.next());
    }
}
```

A. Tom　　B. null　　C. " "　　D. 以上结果都错误

（4）实现 Set 接口的类是（　　）。

A. ArrayList　　B. HashSet　　C. Collection　　D. HashTable

### 2. 填空题

（1）________中的每个对象都由一个特殊的键值对组成。键不能重复，值可以重复。

（2）在程序开发中，经常需要遍历集合中的所有元素。针对这种需求，JDK 专门提供了一个接口 ________。

（3）List 是一个接口，它的实现类有多种，最常用的是________和________。

（4）Set 接口的实现类有多种，最常用的是________和________。

（5）Set 接口中没有重复对象，List 接口中可以________重复对象。

### 3. 编程题

（1）从键盘输入 5 位学生的信息（姓名、语文成绩、数学成绩、英语成绩），按照总分从高到低输出到控制台。

（2）实现一个名为“集合比较器”的类，该类用于比较两个集合的大小，并返回以下 3 种结果之一。

①两个集合相等（即包含相同的元素）。

②第一个集合大于第二个集合（即第一个集合包含第二个集合的所有元素，并且还有

其他元素）。

③第一个集合小于第二个集合（即第一个集合包含的元素少于第二个集合的元素）。

要求如下。

（1）使用 Java 的集合框架（例如 ArrayList、HashSet 等）实现该类。

（2）实现一个静态方法用于比较两个集合的大小，返回一个 String 型的比较结果。

（3）在该类中添加一个构造函数，用于初始化两个需要比较的集合。

第　章

# 字符串类、Lambda表达式与Stream

——凡事勤则易，凡事惰则难

# 7.1 字符串类

## 学习目标

（1）理解字符串的解析。

（2）掌握字符串类的初始化与使用方法。

（3）掌握 String 类、StringBuilder 类、StringBuffer 类的相关操作方法。

## 相关知识

### 7.1.1 String 类

#### 一、String 类对象的初始化

在使用 String 类之前，需要对 String 类对象进行初始化。在 Java 中可以通过以下两种方式对 String 类对象进行初始化。

（1）使用字符串常量初始化一个 String 类对象，例如：

```
String s = "Hello Java"
```

（2）使用 String 类的构造方法初始化 String 类对象。String 类的构造方法如表 7-1-1 所示。

表 7-1-1 String 类的构造方法

| 方法 | 功能描述 |
| --- | --- |
| String( ) | 创建一个内容为空的字符串 |
| String(String value) | 根据指定字符串创建对象 |
| String(char[ ] value) | 根据指定字符数组创建对象 |

【例 7-1-1】根据传入参数的不同，调用不同的 String 类的构造方法来初始化字符串。代码如下：

```
1  public class P7_1_1 {
2      public static void main(String[ ] args){
3          String str1 = new String(); //创建一个空的字符串
4          //创建一个内容为 This is first String 的字符串
5          String str2 = new String("This is first String");
6          //创建一个内容为字符数组的字符串
7          char [ ] array = new char [ ]{'a', 'b', 'c', 'd'};
```

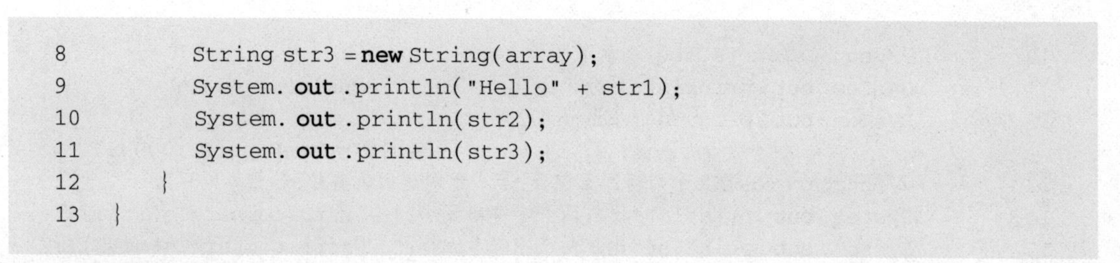

```
8          String str3 = new String(array);
9          System. out .println("Hello" + str1);
10         System. out .println(str2);
11         System. out .println(str3);
12     }
13 }
```

运行结果如图 7－1－1 所示。

```
Hello
This is first String
abcd
```

图 7－1－1　运行结果

## 二、String 类的常用方法

（1）字符串比较方法，如表 7－1－2 所示。

表 7－1－2　字符串比较方法

| 方法 | 功能描述 |
|---|---|
| == | 根据字符串的内存地址是否相同来判断两个字符串是否相等 |
| boolean equals(Object anObject) | 根据字符串的内容是否相同来判断两个字符串是否相等，区分大小写 |
| boolean equalsIgnoreCase(String anotherString) | 根据字符串的内容是否相同来判断两个字符串是否相等，不区分大小写 |
| int compareTo(String anotherString) | 判断两个字符串的 ASCII 码的大小关系，若大于则返回正整数，若相等则返回 0，若小于则返回负整数 |
| int compareToIgnoreCase(String anotherString) | 进行字符串大小关系比较，不区分大小写 |

**【例 7－1－2】** 代码如下：

```
1  public class P7_1_2 {
2      public static void main(String[] args) {
3          String str1 = new String("good");
4          String str2 = new String("good");
5          String str3 = "Good";
6          //==比较两个对象的内存地址是否相等
7          System. out .println("两个值的比较:" +(str1 ==(str2)));
8          //.equal:比较两个字符串的值是否相等
9          System. out .println("两个值的比较:" +(str1.equals(str2)));
```

```
10          //.equalsIgnoreCase:表示不区分大小写
11          System.out.println("区分大小写:"+(str1.equals(str3)));
12          System.out.println("不区分大小写:"
                                  +(str1.equalsIgnoreCase(str3)));
13          //.compareTo:若大于则返回正整数,若小于则返回负整数,若相等则返回0
14          System.out.println("比较两个字符串大小:"+(str1.compareTo(str3)));
15          System.out.println("比较两个字符串大小:"+(str1.compareTo(str2)));
16          //.compareToIgnoreCase:表示不区分大小写
17          System.out.println("不区分大小写比较两个字符串大小:"
                                  +(str1.compareToIgnoreCase(str3)));
18      }
19  }
```

运行结果如图 7-1-2 所示。

```
两个值的比较：false
两个值的比较：true
区分大小写：false
不区分大小写：true
比较两个字符串大小：32
比较两个字符串大小：0
不区分大小写比较两个字符串大小：0
```

图 7-1-2 运行结果

（2）字符串查找方法，如表 7-1-3 所示。

表 7-1-3 字符串查找方法

| 方法 | 功能描述 |
|---|---|
| int indexOf(String str) | 返回指定子字符串在此字符串中第一次出现的位置索引，若找不到则返回 -1 |
| int indexOf(String str, int fromIndex) | 从指定位置开始查找指定子字符串在此字符串中第一次出现的位置索引，若找不到则返回 -1 |
| int lastIndexOf(String str) | 返回指定子字符串在此字符串中最后一次出现的位置索引，若找不到则返回 -1 |
| char charAt(int index) | 返回字符串中索引 index 位置处的字符，其中，索引 index 的取值范围是 0~(字符串长度 -1) |
| boolean endsWith(String suffix) | 判断此字符串是否以指定字符串结尾 |
| boolean startsWith(String prefix) | 判断此字符串是否以指定字符串开头 |
| boolean startsWith(String prefix, int toffset) | 从指定索引位置处判断此字符串是否以指定字符串开头 |
| boolean contains(CharSequence cs) | 判断此字符串是否包含指定字符序列 |

【例 7 -1 -3】代码如下：

```
1  public class P7_1_3 {
2      public static void main(String[] args) {
3          String str = "青葡萄,紫葡萄，吃葡萄不吐葡萄皮,不吃葡萄倒吐葡萄皮";
4          //contain:字符串是否包含子字符串,若包含则返回 true,否则返回 false
5          System.out.println("包含 contain:" + str.contains("葡萄"));
6          //查找子串出现的位置
7          System.out.println("子串出现位置 indexOf:" + str.indexOf("葡萄"));
8          System.out.println("子串最后一次出现位置 lastIndexOf:"
                                            + str.lastIndexOf("葡萄"));
9          System.out.println("从指定位置开始查找子串出现的位置 indexOf:"
                                            + str.indexOf("葡萄",10));
10          //charAt:返回指定位置处的字符
11          System.out.println("返回指定位置字符 charAt:" + str.charAt(4));
12          //startsWith、endsWith:判断是否以指定字符串开头或者结尾
13          System.out.println("匹配开头 startsWith:" + str.startsWith("皮"));
14          System.out.println("匹配结尾 endsWith:" + str.endsWith("皮"));
15      }
16  }
```

运行结果如图 7 -1 -3 所示。

```
包含contain：true
子串出现的位置indexOf:1
子串最后一次出现的位置lastIndexOf:24
从指定位置开始查找子串出现的位置indexOf:10
返回指定位置字符charAt：紫
匹配开头startsWith:false
匹配结尾endsWith:true
```

图 7 -1 -3　运行结果

（3）字符串截取方法，如表 7 -1 -4 所示。

表 7 -1 -4　字符串截取方法

| 方法 | 功能描述 |
| --- | --- |
| String substring( int beginIndex) | 从指定位置到结尾截取字符串 |
| String substring( int beginIndex, int endIndex) | 从指定起始位置到结束位置截取字符串，不包含 endIndex 所在位置处的字符 |

（4）字符串拆分方法，如表 7 -1 -5 所示。

表 7 -1 -5　字符串拆分方法

| 方法 | 功能描述 |
| --- | --- |
| String[ ] split( String regex) | 将字符串按照指定子字符串作为分隔符进行拆分 |
| String[ ] split( String regex, int limit) | 将字符串按照指定子字符串作为分隔符拆分为指定长度的数组 |

【例 7-1-4】代码如下：

```
1  public class P7_1_4 {
2      public static void main(String[] args) {
3          String str = "青葡萄,紫葡萄,吃葡萄不吐葡萄皮,不吃葡萄倒吐葡萄皮";
4          //substring:截取字符串,注意有两个参数时,截取字符串包头不包尾
5          System.out.println("截取字符串到结尾:" + str.substring(4));
6          System.out.println("指定始终位置截取字符串:" + str.substring(4,6));
7          //split:把字符串以","为分隔符拆分为字符串数组
8          String[] strArray1 = str.split(",");
9          for (String s : strArray1) {
10             System.out.println(s);
11         }
12         //split:把字符串以","为分隔符拆分为 2 个字符串
13         String[] strArray2 = str.split(",",2);
14         for (String s : strArray2) {
15             System.out.println(s);
16         }
17     }
18 }
```

运行结果如图 7-1-4 所示。

```
截取字符串到结尾:紫葡萄,吃葡萄不吐葡萄皮,不吃葡萄倒吐葡萄皮
指定始终位置截取字符串:紫葡
青葡萄
紫葡萄
吃葡萄不吐葡萄皮
不吃葡萄倒吐葡萄皮
青葡萄
紫葡萄,吃葡萄不吐葡萄皮,不吃葡萄倒吐葡萄皮
```

图 7-1-4　运行结果

（5）其他字符串操作方法，如表 7-1-6 所示。

表 7-1-6　其他字符串操作方法

| 方法 | 功能描述 |
| --- | --- |
| String concat(String str) | 连接两个字符串 |
| int length() | 返回字符串长度 |
| replace(CharSequence target, CharSequence replacement) | 替换所有匹配的子字符串 |
| String replaceAll(String regex, String replacement) | 替换所有匹配的子字符串 |
| String replaceFirst(String regex, String replacement) | 替换匹配的首个子字符串 |
| boolean isBlank() | 判断字符串是否为空或者是否由空白字符组成（JDK11 提供） |

续表

| 方法 | 功能描述 |
| --- | --- |
| boolean isEmpty( ) | 判断是否为空字符串（即长度为0） |
| String trim( ) | 删除前导和尾随空格 |
| String strip( ) | 同trim( )方法，但是保留Unicode空字符（JDK11提供） |

【例7-1-5】代码如下：

```
1  public class P7_1_5 {
2      public static void main(String[] args) {
3          String str = "青葡萄,紫葡萄,吃葡萄不吐葡萄皮,不吃葡萄倒吐葡萄皮";
4          System.out.println("长度为:" + str.length());
5          String s = "abcd";
6          System.out.println("连接concat:" + s.concat(str));
7          System.out.println("替换replace:" + str.replace("葡萄", "grape"));
8          String s1 = "   a b c d\u0000 \u0000";
9          System.out.println("去空strip:" + s1.strip().length());
10         System.out.println("去空trim:" + s1.trim().length());
11     }
12 }
```

运行结果如图7-1-5所示。

```
长度为：26
连接concat:abcd青葡萄,紫葡萄,吃葡萄不吐葡萄皮,不吃葡萄倒吐葡萄皮
替换replace:青grape,紫grape,吃grape不吐grape皮,不吃grape倒吐grape皮
去空strip:9
去空trim:7
```

图7-1-5　运行结果

## 7.1.2　StringBuffer类

String类定义的内容和长度是不可变的，Java提供了一个字符序列可变的类StringBuffer。在操作字符串时，如果该字符串仅用于表示数据类型，则定义为String类即可；如果需要对字符串进行增、删等操作，则使用StringBuffer类。

### 一、StringBuffer类的初始化

StringBuffer类的初始化与String类相同，但实例化只能通过调用构造方法实现，不允许使用“=”进行赋值。StringBuffer类的构造方法如表7-1-7所示。

表 7-1-7　StringBuffer 类的构造方法

| 方法 | 功能描述 |
| --- | --- |
| StringBuffer( ) | 创建一个内容为空的字符串缓冲区，初始长度为 16 个字符 |
| StringBuffer(String str) | 创建一个内容为 str 的字符串缓冲区 |

**【例 7-1-6】** 根据传入参数的不同，调用不同的 StringBuffer 类的构造方法来初始化字符串。代码如下：

```
public class P7_1_6 {
    public static void main(String[] args) {
        StringBuffer sb = new StringBuffer("This is StringBuffer String!");
        sb = "I am a student"; //报错
    }
}
```

## 二、StringBuffer 类的常用方法

由于 StringBuffer 类具有可变性，所以它可以直接对原字符串进行增、删、替换等操作，而不会产生新的 StringBuffer 类对象。StringBuffer 类的常用方法如表 7-1-8 所示。

表 7-1-8　StringBuffer 类的常用方法

| 方法 | 功能描述 |
| --- | --- |
| StringBuffer append(char c) | 追加字符到 StringBuffer 对象中 |
| StringBuffer insert(int offset, String str) | 在字符串中的 offset 位置前插入字符串 str |
| StringBuffer deleteCharAt(int index) | 删除指定位置的字符 |
| StringBuffer delete(int start, int end) | 删除 StringBuffer 对象中指定范围的字符或字符串序列，不包含 end 所在位置处的字符 |
| StringBuffer replace(int start, int end, String s) | 在 StringBuffer 对象中替换指定的字符或字符串序列，不包含 end 所在位置处的字符 |
| String substring(int beginIndex) | 从指定位置到结尾截取出字符串 |
| String substring(int beginIndex, int endIndex) | 从指定起始位置到结束位置截取字符串，不包含 endIndex 所在位置处的字符 |
| void setCharAt(int index, char ch) | 修改指定索引 index 位置处的字符序列 |
| String toString( ) | 返回 StringBuffer 缓冲区中的字符串 |
| StringBuffer reverse( ) | 将此字符序列用其反转形式替换 |

（1）append( )方法和 insert( )方法的区别如下：二者都用于添加字符，区别在于 append( )方法只能将字符添加到缓冲区的末尾，而 insert( )方法则可以在任意位置添加字符。

（2）replace( )方法在 String 类与 StringBuffer 类中的使用方法不同。同名方法 replace( )在两个类中的参数不同，功能也不同。它在 String 类中的功能是查找指定字符串并以新字符串替换它；在 StringBuffer 类中的功能则是将指定范围的字符序列替换为新字符串。

（3）String 类对象可以使用操作符“+”实现字符串连接，StringBuffer 类对象则不能。

**【例 7-1-7】**代码如下：

```
1  public class P7_1_7 {
2      public static void main(String[] args) {
3          StringBuffer sb = new StringBuffer();
4          //直接在原字符串中操作,不生成新字符串
5          sb.append("good morning");
6          System.out.println("追加 append:" + sb);
7          sb.insert(0, "hello,");
8          System.out.println("插入 insert:" + sb);
9          //删除字符,不包括下标为 10 的字符
10         sb.delete(6, 10);
11         System.out.println("删除 delete:" + sb);
12         //替换字符,不包括下标为 5 的字符
13         sb.replace(0, 5, "Hi");
14         System.out.println("替换 replace:" + sb);
15         sb.reverse();
16         System.out.println("逆序 reverse:" + sb);
17     }
18 }
```

运行结果如图 7-1-6 所示。

```
追加append:good morning
插入insert:hello,good morning
删除delete:hello, morning
替换replace:Hi, morning
逆序reverse:gninrom ,iH
```

图 7-1-6　运行结果

### 7.1.3　StringBuilder 类

StringBuilder 类是 JDK 1.5 新增的类，它也是一个可变字符串类。StringBuilder 类与 StringBuffer 类的功能和方法基本相同，不同的是 StringBuffer 类是线程安全的，而 StringBuilder 类由于没有实现线程安全，所以性能高于 StringBuffer 类。因此，在一般情况下，创建一个内容可变的字符串对象，优先考虑使用 StringBuilder 类。

**案例**

在用户信息管理系统中如何实现以身份证号码为用户名的注册功能？如何检测身份证号码的正确性？如何实现密码的核验？

【分析】

(1) 身份证号码验证：测试字符串长度是否为 18、内容是否为阿拉伯数字或者 X。

(2) 密码核验：测试密码是否为指定长度、两次输入密码是否一致。

(3) 相应提示信息。

### 1. 窗体设计

窗体设计如图 7－1－7 所示。

图 7－1－7　窗体设计

### 2. 功能实现

代码如下：

```
public class Register extends JFrame {
    private JPanel contentPane;
    private JTextField textField;
    private JPasswordField passwordField , passwordField_1;
    JLabel label_1 ,label_5;
    public static void main(String[] args) {
        ……(略)
    }

    public Register() {
          setTitle("注册");
          ……(略)
          JButton button  = new JButton("去空");
          button.addActionListener(new ActionListener() {
              public void actionPerformed(ActionEvent e) {
                  String s = textField.getText();
                  String s1 = s.replace(" ", "");
                  textField.setText(s1);
              }
        });
        button.setBounds(498, 35, 99, 29);
        contentPane.add(button);
```

```
        JButton button_1 = new JButton("检查");
        button_1.addActionListener(new ActionListener() {
            public void actionPerformed(ActionEvent e) {
              String s = textField.getText();
              System.out.println(s.length());
              if (s.length() ==18){ //1.判断字符串的长度是不是18位
                  System.out.println("您的身份证号码是:" + s);
                  label_1.setText("您的身份证号码是:" + s);
              }
              else {
                  System.out.println("身份证号码位数错误");
                  label_1.setText("身份证号码位数错误");
              }
            }
          });
          button_1.setBounds(602, 35, 99, 29);
          contentPane.add(button_1);
          ……(略)
          JButton button_2 = new JButton("注册");
          button_2.addActionListener(new ActionListener() {
              public void actionPerformed(ActionEvent e) {
                  String psw = new String(passwordField.getPassword());
                  String psw1 = new String(passwordField_1.getPassword());
                  System.out.println("psw = " + psw);
                  //1.判断密码是否只包含数字
                  //2.判断密码是不是6位
                  //3.判断两个密码框中的值是不是相同
                  //4.若以上条件都满足,则登录成功
                  String regex = "^[0-9]+$";
                  if (!psw.matches(regex)){
                      label_5.setText("密码只能包含数字");
                  }else if (psw.length() >6){
                        label_5.setText("密码超过6位");
                  }else if (!psw.equals(psw1)){
                        label_5.setText("两次密码不同");
                  }else {
                        label_5.setText("注册成功");
                        Main m1 = new Main(); //生成一个主窗体
                  m1.setVisible(true); //把主窗体显示出来
                      Register.this.dispose(); //把当前注册窗体关闭
                  }
              }
        });
        ……(略)
    }
}
```

## 任务工单

| 任务名称 | 图书信息管理系统——图书检索 | | | | |
| --- | --- | --- | --- | --- | --- |
| 班级 | | 学号 | | 姓名 | |

任务要求



（1）定义图书列表，代码如下：

```
List<Book> books = new ArrayList<Book>();
books.add(new Book("001","Java 程序设计","王强","清华大学出版社",50,"在库"));
books.add(new Book("002","Java 案例教程","王新","北京大学出版社",58,"在库"));
books.add(new Book("003","Java 程序设计从入门到精通","周北平","清华大学出版社",46,"在库"));
books.add(new Book("004","C 语言程序设计","何贝","高等教育出版社",38,"在库"));
books.add(new Book("005","C 语言案例教程","周华金","高等教育出版社",40,"在库"));
books.add(new Book("006","Python 程序设计","林琳","北京大学出版社",50,"在库"));
```

（2）在图书列表中查询包含窗体中检索关键字的图书，并在多行文本框中展示。

**续表**

<table>
<tr><td>任务名称</td><td colspan="5">图书信息管理系统——图书检索</td></tr>
<tr><td>班级</td><td></td><td>学号</td><td></td><td>姓名</td><td></td></tr>
<tr><td>任务实现</td><td colspan="5">（实现检索按钮功能）<br><br>JButton btnNewButton = new JButton("检索");<br>        btnNewButton.addActionListener(new ActionListener() {<br>            public void actionPerformed(ActionEvent e)<br><br><br><br>        }<br>    });</td></tr>
<tr><td>异常及<br>待解决<br>问题记录</td><td colspan="5"></td></tr>
</table>

# 7.2 Lambda 表达式

## 学习目标

（1）熟悉 Lambda 表达式的语法格式。

（2）掌握 Lambda 表达式的使用方法。

## 相关知识

### 7.2.1 Lambda 表达式的概念

Java 8 的一个亮点是引入了 Lambda 表达式，Lambda 表达式可以代替匿名内部类实现接口，Lambda 表达式的本质就是匿名函数。

#### 一、函数接口

如果一个接口只包含一个抽象方法，那么这个接口就是函数接口，一般用@ FunctionalInterface 标注函数接口（也可省略）。例如：

```
1  @ FunctionalInterface
2  interface Calculate{
3          int operate(int a,int b);
4  }
```

Java API 中符合函数接口的例子不少，例如：

```
(1)public interface Runnable{void run();} //线程要实现的接口
(2)public interface ActionListener{void actionPerformed(ActionEvent e);} //动作
监听要实现的接口
```

以往采用匿名内嵌类实现接口是最简单的方法（如下），而 Lambda 表达式使函数接口的实现更加简便。

```
1  Calculate c = new Calculate() {
2      @ Override
3      public int operate(int a, int b) {
4          return a + b;
5      }
6  }
```

#### 二、Lambda 表达式的使用

Lambda 表达式的本质是匿名函数，它由 3 个部分组成：参数列表、箭号（或称为 goes to）、

方法体，具体语法格式如下：

```
(参数列表)->{方法体}
```

【例7-2-1】代码如下：

```
1  //定义6个接口
2  interface It1{
3      void test();//无参、无返回值的抽象方法
4  }
5  interface It2{
6      void test(int a); //有1个参数、无返回值的抽象方法
7  }
8  interface It3{
9      void test(int a,int b); //有2个参数、无返回值的抽象方法
10 }
11 interface It4{
12     int test();//无参、有返回值的抽象方法
13 }
14 interface It5{
15     int test(int a); //有1个参数、有返回值的抽象方法
16 }
17 interface It6{
18     int test(int a,int b); //有2个参数、有返回值的抽象方法
19 }
20
21 public class P7_2_1 {
22     public static void main(String[] args) {
23         //Lambda 表达式实现6个接口
24         It1 it1 = () ->{
25             System. out .println("无参无返回值");
26         };
27         It2 it2 = (int a) ->{
28             System. out .println("1个参数无返回值:" +a);
29         };
30         It3 it3 = (int a,int b) ->{
31             System. out .println("2个参数无返回值:" +a +b);
32         };
33         It4 it4 = () ->{
34             return 0;
35         };
36         It5 it5 = (int a) ->{
37             return a;
38         };
39         It6 it6 = (int a,int b) ->{
40             return a +b;
41         };
42         it1.test();
43         it2.test(2);
```

```
44          it3.test(2,3);
45          System.out.println("无参有返回值:" +it4.test());
46          System.out.println("1 个参数有返回值:" +it5.test(2));
47          System.out.println("2 个参数有返回值:" +it6.test(2,3));
48      }
49  }
```

运行结果如图 7-2-1 所示。

```
无参无返回值
1个参数无返回值：2
2个参数无返回值：23
无参有返回值:0
1个参数有返回值:2
2个参数有返回值:5
```

图 7-2-1　运行结果

**1. Lambda 表达式的参数**

（1）参数可以是 0 个、1 个、2 个或者多个。

（2）无参数时小括号不能省略，例如【例 7-2-1】中的第 24 行代码。

（3）参数可以显式声明类型，也可以由编译器根据上下文推断，例如【例 7-2-1】中的第 27 行代码，也可写成“It2 it2 =(a) ->{ }”，即（int a）等同于（a）。

（4）参数用小括号括起来，参数间用逗号隔开，例如【例 7-2-1】中的第 30 行代码。

（5）当参数有且仅有 1 个时，如果不显式声明类型，则不必使用小括号，例如【例 7-2-1】中的第 27 行代码，也可写成“It2 it2 = a -> { }”，即（a）等同于 a。

**2. Lambda 表达式的方法体**

（1）方法体可以包含 0 条、1 条或多条语句。

（2）若方法体中只有 1 条语句，则大括号可不用写，且返回值类型要与匿名函数的返回值类型相同，例如【例 7-2-1】中的所有 Lambda 表达式可精简如下：

```
It1 it1 =() ->System.out.println("无参无返回值");
It2 it2 =a ->System.out.println("1 个参数无返回值:" +a);
It3 it3 =(a,b) ->System.out.println("2 个参数无返回值:" +a +b);
It4 it4 =() ->return 0;
It5 it5 =a ->return a;
It6 it6 =(a,b) ->return a +b;
```

（3）若方法体中有 1 条以上的语句，则必须包含在大括号（代码块）中，且返回值类型要与匿名函数的返回值类型相同。

综上，引入采用 Lambda 表达式，Calculate 接口的实现方法可简化为【例 7-2-2】。

**【例 7-2-2】**代码如下：

```
1  interface Calculate{
2      int operate(int a,int b);
```

```
3  }
4  public class P7_2_2 {
5      public static void main(String[] args) {
6      //1行Lambda表达式替代6行内部函数,精简代码
7      Calculate cal01 = (a,b) -> a + b;
8      System.out.println(cal01.operate(4, 5));
9    }
10}
```

### 7.2.2　泛型与Lambda表达式

在Lambda表达式中，编译器会根据泛型参数推断出Lambda表达式的参数类型。

**【例7-2-3】** Calculate接口及其实现方法。代码如下：

```
1  interface Calculate<T>{
2      T operate(T a,T b);
3  }
4
5  public class P7_2_3 {
6      public static void main(String[] args) {
7          Calculate<String> s1 = (x,y) -> x + y;
8          Calculate<Integer> s2 = (x,y) -> x + y;
9          System.out.println("s1 = " + s1.operate("1", "2"));
10         System.out.println("s2 = " + s2.operate(1, 2));
11     }
12 }
```

【说明】

程序第1行，Calculate接口引入泛型T，第7、8行的Lambda表达式相同，但是泛型实参变化导致调用结果不同，第7行实现两个字符串连接，第8行实现两个整数相加。

### 7.2.3　Lambda表达式的方法引用

#### 一、通用方法的引用

在程序开发中，常出现多个Lambda表达式实现相同功能的情况，这时可以将Lambda表达式封装成通用方法，以便后期维护。这可以采用Lambda表达式的方法引用实现，其语法格式如下：

```
对象::方法 或者 类::方法(静态方法)
```

**【例7-2-4】** 代码如下：

```
1  interface Calculate{
2      int operate(int a,int b);
```

```
3  }
4  public class P7_2_4 {
5      public static void main(String[] args) {
6          //应用1:定义 Lambda 表达式实现 Calculate 接口的方法
7          Calculate cal01 = (a,b) -> a + b;
8          //应用2:定义 Lambda 表达式实现 Calculate 接口的方法
9          Calculate cal02 = (a,b) -> a + b;
10     }
11 }
```

【说明】

程序第7行和第9行，两个应用分别定义了实现接口 Calculate 的方法，当功能发生变化时，需要修改两个方法的实现代码，为了简化维护工作，将实现接口的方法定义为通用方法，采用 Lambda 表达式的方法引用来调用，例如【例7-2-5】。

**【例7-2-5】** 代码如下：

```
1  interface Calculate{
2      int operate(int a,int b);
3  }
4  public class P7_2_5 {
5      //定义通用方法
6      int operateAll(int a,int b) {
7          return a + b;
8      }
9      public static void main(String[] args) {
10         //实例化 P6_11
11         P7_2_5 p = new P7_2_5();
12         //应用1:使用 Lambda 表达式的方法引用调用 operateAll 方法
13         Calculate cal01 = p::operateAll;
14         //应用2:使用 Lambda 表达式的方法引用调用 operateAll 方法
15         Calculate cal02 = p::operateAll;
16     }
17 }
```

【说明】

程序第6行定义了通用方法 operateAll(int a, int b)，第13、15行使用 Lambda 表达式的方法引用调用 operateAll()方法实现 Calculate 接口的 operate(int a, int b) 方法。当功能变化时，只需要修改 operateAll(int a, int b) 方法即可。若 operateAll(int a, int b) 方法为静态方法，则可以直接使用 P6_11:: operateAll 来实现引用，具体实现如下：

```
9      public static void main(String[] args) {
10          //应用1:使用 Lambda 表达式的方法引用调用 operateAll 方法
11         Calculate cal01 = P6_11::operateAll;
12         //应用2:使用 Lambda 表达式的方法引用调用 operateAll 方法
13         Calculate cal02 = P6_11::operateAll;
```

```
14         }
15  }
```

## 二、构造方法的引用

如果函数接口的实现可以直接在一个类的构造方法中实现，那么就可以使用构造方法的引用，其语法格式如下：

```
类名::new
```

【例7-2-6】使用 Lambda 表达式，但不使用构造方法的引用，实现函数接口 PS1、PS2。代码如下：

```
1  public class Person {
2      private String name;
3      private int age;
4      //省略 get()、set()方法、构造方法、toString()方法
5  }
6  //函数接口 1
7  interface PS1{
8          //获取一个空的人类对象的抽象方法
9          Person getPerson();
10 }
11 //函数接口 2
12 interface PS2{
13         //获取一个带全参的人类对象的抽象方法
14         Person getPerson(String name,int age);
15 }
16 public class P7_2_6 {
17         public static void main(String[] args) {
18         //使用 Lambda 表达式构造方法的引用实现 PS1 接口
19         PS1 ps1 = Person::new ;
20         //调用无参构造方法
21         System. out .println(ps1.getPerson());
22         //使用 Lambda 表达式构造方法的引用实现 PS2 接口
23         PS2 ps2 = Person::new ;
24         //调用带全参构造方法
25         System. out .println(ps2.getPerson("杨洋",20));
26         }
27 }
```

运行结果如图7-2-2所示。

```
Person [name=null, age=0]
Person [name=杨洋, age=20]
```

图7-2-2　运行结果

【例7-2-7】代码如下：

```
1  public class P7_2_7{
2      public static void main(String[] args){
3          int r =10;
4          double area = Math.PI * Math.pow(r,2);
5          System.out.println("半径为 10 的圆形面积是:" + area);
6      }
7  }
```

运行结果如图 7-2-3 所示。

```
半径为10的圆形面积是:314.1592653589793
```

图 7-2-3　运行结果

### 7.2.4　系统内置四大函数接口

Java 8 提供了一系列内置的函数接口，存储在 java. util. function 包中，这些函数接口提供了某类功能的操作规范，通过这些函数接口类型的引用变量可以引用符合其要求的实际函数。常见的函数接口如表 7-2-1 所示。

表 7-2-1　常见的函数接口

| 函数接口 | 参数类型 | 返回类型 | 说明 |
|---|---|---|---|
| Consumer <T> 消费型接口 | T | void | void accept(T t)：对类型为 T 的对象应用操作 |
| Supplier <T> 供给型接口 | 无 | T | T get()：返回类型为 T 的对象 |
| Function <T，R> 函数型接口 | T | R | R apply(T t)：对类型为 T 的对象应用操作，并返回类型为 R 的对象 |
| Predicate <T> 断言型接口 | T | boolean | boolean test(T t)：确定类型为 T 的对象是否满足条件，并返回类型为 boolean 的对象 |

【例 7-2-8】代码如下：

```
1  import java.util.function.*;
2  public class P7_2_8 {
3      public static void main(String[] args) {
4          //Consumer<T>消费型接口:void accept(T t)
5          Consumer<Double> consumer = a -> System.out.println("花了:" + a + "元");
6          consumer.accept(10.0);
7          //Supplier<T>供给型接口:T get()
8          Supplier<String> supplier = () -> {
9              Person p = new Person("杨洋",20);
```

```
10              return p.getName();
11          }
12          System.out.println(supplier.get());
13          //Function<T,R>函数型接口:R apply(T t)
14          Function<String, String> function = s1 -> s1.toUpperCase();
15          System.out.println(function.apply("java"));
16          //Predicate<T>断言型接口:boolean test(T t)
17          Predicate<Person> predicate = p -> p.getAge() > 30 && p.getAge() < 60;
18          System.out.println(predicate.test(new Person("杨洋",20)));
19      }
20  }
```

运行结果如图7-2-4所示。

```
花了:10.0元
杨洋
JAVA
false
```

图7-2-4 运行结果

**案例**

在社区管理系统中，如何实现对所有居民按照年龄进行排序的功能?

【分析】

(1) 居民数据组织：定义封装类Resident。

(2) 社区所有居民的数据组织：定义一个存放Resident的集合。

(3) 使用集合的sort()方法实现按照年龄排序的功能，查看sort()方法的定义如下：

```
  @SuppressWarnings({"unchecked", "rawtypes"})
 default void sort(Comparator<? super E> c) {
    Object[] a = this.toArray();
    Arrays.sort(a, (Comparator) c);
    ListIterator<E> i = this.listIterator();
    for (Object e : a) {
        i.next();
        i.set((E) e);
    }
 }
```

(4) 查看函数接口Comparator如下：

```
@FunctionalInterface
   public interface Comparator<T> {
   int compare(T o1, T o2);
   ……(略)
}
```

Comparator接口有两个参数，因此Lambda表达式必须提供两个参数，即表达式为(n1, n2) -> { n1.getAge() - n2.getAge() }。

### 1. 窗体设计

窗体设计如图 7 – 2 – 5 所示。

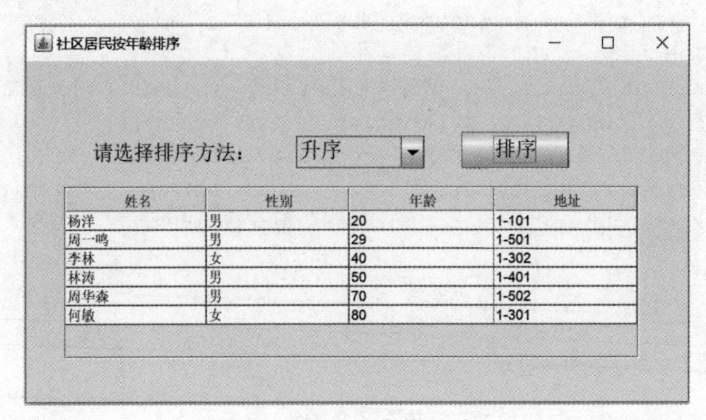


图 7 – 2 – 5　窗体设计

### 2. 功能实现

代码如下:

```
public class P7 extends JFrame {
      private static final long serialVersionUID = 1L;
      private JPanel contentPane;
      List <Resident > list;
      private JTable table;

      public static void main(String[] args) {
      ……(略)
}
public P7() {
      setTitle("社区居民按年龄排序");
      ……(略)

      list =new ArrayList <Resident >();
      list.add(new Resident("杨洋","男",20,"1 -101"));
      list.add(new Resident("周一鸣","男",29,"1 -501"));
      list.add(new Resident("林涛","男",50,"1 -401"));
      list.add(new Resident("何敏","女",80,"1 -301"));
      list.add(new Resident("周华森","男",70,"1 -502"));
      list.add(new Resident("李林","女",40,"1 -302"));

      JButton btnNewButton  = new JButton("排序");
      btnNewButton.addActionListener(new ActionListener() {
         public void actionPerformed(ActionEvent e) {
           String s =comboBox.getSelectedItem().toString();
           if (s.equals("升序"))
               list.sort((n1,n2) ->n1.getAge() -n2.getAge());
         else
               list.sort((n1,n2) ->n2.getAge() -n1.getAge());
         System. out .println(list);
```

```
            Vector data = new Vector();
            for (Resident r : list) {
                Vector row = new Vector();
                row.add(r.getName());
                row.add(r.getSex());
                row.add(r.getAge());
                row.add(r.getAddr());
                data.add(row);
            }
            Vector columns = new Vector();
            columns.add("姓名");
            columns.add("性别");
            columns.add("年龄");
            columns.add("地址");
            DefaultTableModel m = new DefaultTableModel();
            m.setDataVector(data, columns);
            table.setModel(m);
        }
    });
    ......(略)
  }
}
```

## 任务工单

| 任务名称 | 图书信息管理系统——删除图书信息 | | | | |
| --- | --- | --- | --- | --- | --- |
| 班级 | | 学号 | | 姓名 | |

**任务要求**

删除图书信息（Lambda）

请选择图书编号： 001 删除

Book [bookID=002, bookName=Java案例教程, writer=王新, press=北京大学出版社, price=58.0]
Book [bookID=003, bookName=Java程序设计从入门到精通, writer=周北平, press=清华大学出版社, price=46.0]
Book [bookID=004, bookName=C语言程序设计, writer=何贝, press=高等教育出版社, price=38.0]
Book [bookID=005, bookName=C语言案例教程, writer=周华金, press=高等教育出版社, price=40.0]
Book [bookID=006, bookName=Python程序设计, writer=林琳, press=北京大学出版社, price=50.0]

（1）定义图书列表，代码如下：

```
List<Book> books = new ArrayList<Book>();
books.add(new Book("001","Java 程序设计","王强","清华大学出版社",50,"在库"));
books.add(new Book("002","Java 案例教程","王新","北京大学出版社",58,"在库"));
books.add(new Book("003","Java 程序设计从入门到精通","周北平","清华大学出版社",46,"在库"));
books.add(new Book("004","C 语言程序设计","何贝","高等教育出版社",38,"在库"));
books.add(new Book( "005","C 语言案例教程","周华金","高等教育出版社",40,"在库"));
books.add(new Book( "006","Python 程序设计","林琳","北京大学出版社",50,"在库"));
```

（2）要求 JComboBox 显示图书列表中的所有图书编号。

（3）定义 Delete 接口，代码如下：

```
public interface Delete<T> {
List<T> delete(String bid);
}
```

（4）选择图书编号，删除图书列表中的指定图书（使用 Lambda 表达式实现 Delete 接口完成此功能）。

**续表**

<table>
<tr><td>任务名称</td><td colspan="5">图书信息管理系统——删除图书信息</td></tr>
<tr><td>班级</td><td></td><td>学号</td><td></td><td>姓名</td><td></td></tr>
<tr><td>任务实现</td><td colspan="5">（补充下面的程序片段，实现功能）<br>（1）设置 JComboBox 值：<br><br><br><br>（2）删除按钮功能：<br><br><br><br>JButton btnNewButton = new JButton("删除");<br>btnNewButton.addActionListener(new ActionListener() {<br>public void actionPerformed(ActionEvent e) {<br><br><br><br>}<br>});</td></tr>
<tr><td>异常及待解决问题记录</td><td colspan="5"></td></tr>
</table>

# 7.3 Stream

## 学习目标

（1）熟悉 Stream 的工作过程。

（2）掌握创建 Stream 的方法。

（3）掌握 Stream 的中间操作方法。

（4）掌握 Stream 的最终操作方法。

## 相关知识

Stream 是 Java 8 引入的一种新的数据处理方式，是对集合对象的功能增强，它支持类似 SQL 语句的操作，例如过滤、映射、排序等操作，结合 Lambda 表达式提高了编程效率，使代码更加简洁、易读。

### 一、Stream 的工作过程

（1）创建 Stream 对象：为指定数据源创建 Stream 对象，数据源可以是集合、数组、IO 通道等。

（2）中间操作：对 Stream 对象进行一系列转换和操作，包括过滤、映射、排序等。

（3）最终操作：通过聚合、收集、遍历等手段从 Stream 对象中获取结果。

Stream 的工作过程如图 7－3－1 所示。

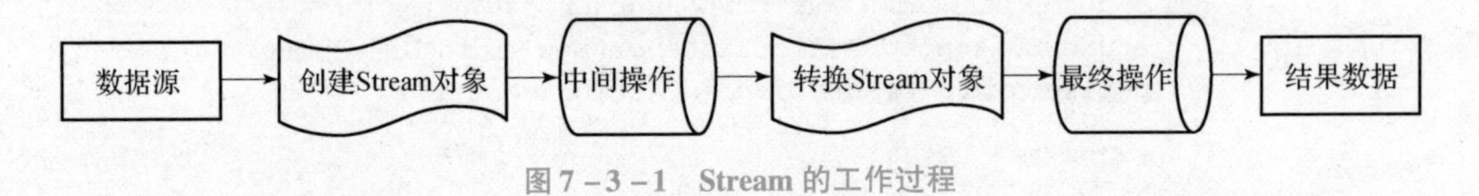


图 7－3－1　Stream 的工作过程

**【例 7－3－1】** 代码如下：

```
1  public class P7_3_1 {
2      public static void main(String[] args) {
3          //创建数据源
4          List<Integer> nums = Arrays.asList(1,null,3,4,null,6);
5          Long n = nums.stream() //创建 Stream 对象
6                  .filter(num -> num != null ) //中间操作
7                  .count(); //最终操作
8          System.out.println(n);
9      }
10 }
```

【说明】

程序获取 List 集合中不为 null 的元素个数。Stream 可以看作高级版的迭代器，区别在于

迭代器只能逐个遍历元素并对其执行某些操作，而 Stream 只对满足条件（如过滤掉 null 值）的元素逐个执行操作（计数）。

## 二、创建 Stream

在 Java 中创建 Stream 的方式主要有 6 种。

### 1. 使用 Collection 集合创建 Stream（最常用）

（1）Collection 接口提供 stream( ) 方法创建串行 Stream。

（2）Collection 接口提供 parallelStream( ) 方法创建并行 Stream。

例如：

```
List<Integer> intList = new ArrayList<Integer>();
intList.add(10);
intList.add(20);
intList.add(30);
intList.add(40);
Stream<Integer> listStream = intList.stream();
listStream.forEach(System.out::println);
```

### 2. 使用 Array 数组创建 Stream

可以使用 Arrays. stream(T array)方法从数组创建 Stream 对象，这里创建的 Stream 是一个原生 Stream（Stream 数据均为整型）。

例如：

```
int[] arr = {10,20,30,40};
IntStream arrStream = Arrays.stream(arr);
arrStream.forEach(System.out::println);
```

### 3. 使用文件创建 Stream

可以使用 Files. line( )方法得到一个 Stream，且得到的每个 Stream 是给定文件中的一行。

例如：

```
try {
        Stream<String> fileStream = Files.lines
            (Paths.get("d:/hello.java"),Charset.defaultCharset());
        fileStream.forEach(a -> System.out.println(a));
} catch (IOException e) {
      e.printStackTrace();
}
```

### 4. 使用 Stream 创建 Stream

（1）Stream<T>接口提供了 of(T)方法来创建 Stream。

（2）Stream<T>接口提供了 empty( )方法来创建不含任何元素的 Stream。

例如：

```
Stream<Integer> stream1 = Stream.of(10,20,30,40);
Stream<Integer> stream2 = Stream.empty();
stream1.forEach(System.out::println);
```

**5. 使用函数创建 Stream**

Stream 接口提供了 2 个创建无限 Stream 的静态方法：iterator( )、generate( )。

（1）iterator( )：接收 2 个参数，第一个为初始化值，第二个为所进行的函数操作。

（2）generate( )：接收 1 个类型为 Supplier 的参数，由它为 Stream 提供值。

例如：

```
Stream<Integer> iteStream = Stream.iterate(0, n->n+1).limit(10);
Stream<Double> genStream = Stream.generate(Math::random).limit(10);
iteStream.forEach(System.out::println);
genStream.forEach(System.out::println);
```

**6. 创建原生 Stream**

基于基础数据类型，Java 提供了 IntStream、LongStream、DoubleStream 等原生 Stream。

例如：

```
IntStream.of(new int[]{10,20,30}).forEach(System.out::println);
IntStream.range(1, 5).forEach(System.out::println);
IntStream.rangeClosed(1, 5).forEach(System.out::println);
//先创建 Stream<Integer>,再转成 IntStream(原生 Stream)
Stream<Integer> stream = Stream.of(10,20,30);
IntStream intStream = stream.mapToInt(Integer::valueOf);
intStream.forEach(System.out::println);
```

## 三、中间操作

Stream 提供了丰富的中间操作，中间操作主要是对数据集进行整理，返回值为结果 Stream，一个 Stream 可以跟随 0 个或者多个中间操作。中间操作包括 filter( )、takeWhile( )、dropWhile( )、map( )、distinct( )、sorted( )、limit( )、skip( )、mapToInt( )、mapToLong( )、mapToDouble( )、flatMap( )、flatMapToInt( )、flatMapToLong( )、flatMapToDouble( )、peek( )、unordered( )。这类方法都采用惰性化设计，调用这类方法并没有开始 Stream 的遍历，真正的遍历必须在最终操作中进行。下面介绍几个常用方法。

**1. filter( )方法、takeWhile( )方法、dorpWhile( )方法**

（1）filter( )方法过滤出满足条件的元素，其语法格式如下：

```
Stream<T> filter(Predicate<? super T> predicate);
```

（2）takeWhile( )方法过滤出函数值首次为 false 之前的元素，其语法格式如下：

```
Stream<T> takeWhile (Predicate<? super T> predicate);
```

（3） dropWhile( )方法过滤出函数值首次为 false 之后的元素，其语法格式如下：

```
Stream<T> dropWhile(Predicate<? super T> predicate);
```

【例 7-3-2】代码如下：

```
1  public class P7_3_2 {
2      public static void main(String[] args) {
3          List<String> s = new ArrayList<String>();
4          s.add("hello");
5          s.add("hi");
6          s.add("how do you do");
7          s.add("how are you");
8          Stream<String> out = s.stream().filter(t ->t.contains("o"));
9          out.forEach(System.out::println);
10         System.out.println("-----------------");
11         Stream<String> out1 = s.stream().takeWhile(t ->t.contains("o"));
12         out1.forEach(System.out::println);
13         System.out.println("-----------------");
14         Stream<String> out2 = s.stream().dropWhile(t ->t.contains("o"));
15         out2.forEach(System.out::println);
16     }
17 }
```

运行结果如图 7-3-2 所示。

```
hello
how do you do
how are you
-----------------
hello
-----------------
hi
how do you do
how are you
```

图 7-3-2　运行结果

【说明】

程序第 8 行 s. stream( )生成 Stream，通过 Stream 的 filter( )方法过滤出包含字符“o”的元素，Lambda 表达式 t -> t. contain("o")设置过滤条件，返回值仍然是一个 Stream。第 11 行通过 Stream 的 takeWhile( )方法首先找出不包含字符“o”的第一个元素为“hi”，结果 Stream 中则包含元素“hi”之前的所有元素：“hello”。第 14 行通过 Stream 的 dropWhile( )方法首先找出不包含字符“o”的第一个元素为“hi”，结果 Stream 中则包含元素“hi”之后的所有元素：“hi”“how do you do”“how are you”。

**2. map( )系列方法**

map( )系列方法接收一个函数 Function<T, R>作为参数，Stream 中的每个元素都按指定函数进行转换。

（1） map( )方法把 Stream 中的每个元素按照指定函数重新映射，元素个数不变，但元素值发生变化。

（2）mapToInt( )、mapToLong( )、mapToDouble( )方法把 Stream 中的每个元素转换成指定数据类型。

（3）flatMap( )、flatMapToInt( )、flatMapToLong( )、flatMapToDouble( )方法把 Stream 中的每个元素进行一对多的拆分，细分为更小的单元，返回的 Stream 元素的个数相应增加。

**【例 7-3-3】** 代码如下：

```
1  public class P7_3_3 {
2      public static void main(String[] args) {
3          List<String> s = Arrays.asList("1","12","123");
4          Stream<Integer> out = s.stream().map(String::length);
5          out.forEach(System.out::println);
6          System.out.println("-----------------");
7          IntStream out1 = s.stream().mapToInt(Integer::parseInt);
8          out1.forEach(System.out::println);
9          System.out.println("-----------------");
10         List<Integer> t1 = Arrays.asList(1,2,3);
11         List<Integer> t2 = Arrays.asList(11,22,33);
12         List<List<Integer>> list = new ArrayList();
13         list.add(t1);
14         list.add(t2);
15         System.out.println("flatMap前:" + list);
16         List<Integer> out2 = list.stream().flatMap(Collection::stream)
                   .collect(Collectors.toList());
17         System.out.println("flatMap后:" + out2);
18     }
19 }
```

运行结果如图 7-3-3 所示。

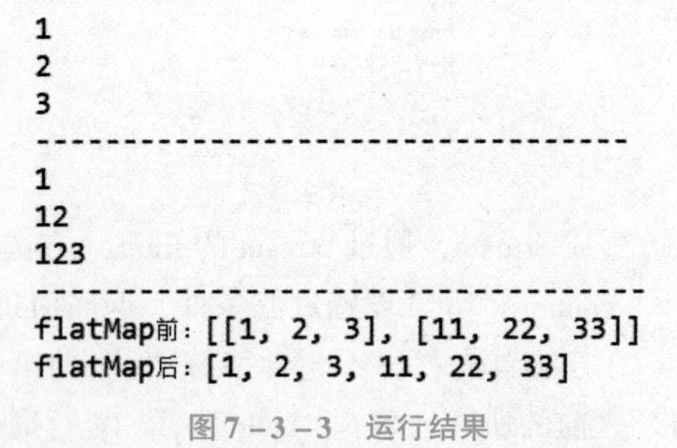
```
1
2
3
-----------------------------------
1
12
123
-----------------------------------
flatMap前：[[1, 2, 3], [11, 22, 33]]
flatMap后：[1, 2, 3, 11, 22, 33]
```

图 7-3-3 运行结果

【说明】

程序第 4 行 s.stream( )生成 Stream，通过 Stream 的 map( )方法指定的映射函数 length( )，返回 Stream 中的每个字符串长度；第 7 行 s.stream( )生成 Stream，通过 Stream 的 mapToInt( )方法指定的映射函数 parseInt( )，将 Stream 中的所有元素转换成 Integer 数据类型；第 12 行定义一个二维数组［［1，2，3］，［11，22，33］］，通过 Stream 的 flatMap( )方法将两个一维数组中的元素逐一分解出来，返回的 Stream 降维为一维数组 out2。

### 3. sorted()方法

（1）sorted()方法用于对Stream中的所有元素进行排序，返回值为排序后的Stream对象，其语法格式如下：

```
①Stream<T> sorted()//无参,对Stream中的元素进行自然排序
②Stream<T> sorted(Comparator<? super T> comparator)//带参,重写Comparator接口的compara()方法指定排序规则,对Stream中的元素按自定义规则进行排序
```

【例7-3-4】代码如下：

```
1  public class P7_3_4 {
2      public static void main(String[] args) {
3          List<Integer> list = Arrays.asList(123,32,98,15,44,201);
4          Stream<Integer> out = list.stream().sorted();
5          out.forEach(System.out::println);
6          System.out.println("------------------");
7          Stream<Integer> out1 = list.stream().sorted((o1,o2)->o2-o1);
8          out1.forEach(System.out::println);
9      }
10 }
```

运行结果如图7-3-4所示。

```
15
32
44
98
123
201
---------------
201
123
98
44
32
15
```

图7-3-4 运行结果

【说明】

程序第4行采用默认升序方式对Stream数据进行排序。第7行重写Comparator接口的compara()方法，定义比较顺序为降序排列。

（2）sorted()方法对封装数据进行排序，为了提高排序效率，可以先将Stream转换为原生Stream，这就减少了数据拆装箱环节。

【例7-3-5】代码如下：

```
1  public class P7_3_5 {
2      public static void main(String[] args) {
```

```
3          List<Integer> list = Arrays.asList(123,32,98,15,44,201);
4          IntStream s = list.stream().mapToInt(Integer::intValue);
5          s.sorted().forEach(k -> System.out.printf("%5d",k));
6      }
7  }
```

运行结果如图 7－3－5 所示。

```
15   32   44   98  123  201
```

图 7－3－5　运行结果

【说明】

程序第 4 行，调用 mapToInt( )方法将 Stream 转为原生 Stream。第 5 行采用默认升序方式对 Stream 数据进行排序并输出。

#### 4. distinct( )方法

distinct( )方法用于删除 Stream 中的重复元素。

**【例 7－3－6】** 代码如下：

```
1  public class P7_3_6 {
2      public static void main(String[] args) {
3          List<Integer> list = Arrays.asList(123,123,32,98,15,44,201);
4      //去除 Stream 中的重复值 123,值保留一份
5      list.stream().distinct().forEach(k -> System.out.printf("%5d", k));
6      }
7  }
```

运行结果如图 7－3－6 所示。

```
15   32   44   98  123  201
```

图 7－3－6　运行结果

### 四、最终操作

最终操作指结束 Stream 的处理方式，同时触发 Stream 的执行。一个 Stream 有且只有一个最终操作，当最终操作执行时，则 Stream 被关闭，无法再执行其他操作。常用的最终操作有 collect( )、count( )、max( )、min( )、forEach( )、reduce( )、toArray( )、anyMatch( )、allMatch( )、noneMatch( )、findFirst( )、findAny( )、forEachOrdered( )等方法。

#### 1. 汇集方法

汇集方法用于完成数据的各类统计。常用的汇集方法如下。

（1）count( )方法用于统计当前 Stream 包含的元素数量，返回一个数值。

（2）sum( )方法用于统计当前 Stream 元素之和，返回元素相应数据的数值。

（3）average( )方法用于统计当前 Stream 元素平均值，返回值为 OptionalDouble 类型，通过 getAsDouble( )方法获取其值。

（4）max( )方法、min( )方法用于统计当前 Stream 中的最大值、最小值。当 max( )、

min( )方法作用于原生 Stream，无须提供参数，返回值为 Stream 元素相应的 Optional 类型。例如，OptionalInt、OptionalDouble 可通过 getAsInt( )、getAsDouble( )方法获取。当 max( )、min( )方法作用于对象 Stream，需要提供比较器作为参数，调用 compareTo( )方法实现数据的比较，返回值为 Optional 类型，可以通过 get( )方法获取。

【例 7-3-7】代码如下：

```
1  public class P7_3_7 {
2      public static void main(String[] args) {
3          List<Integer> list = Arrays.asList(123,32,98,15,44,201);
4          //原生 Stream(原生流)
5          System.out.println("原生流操作:");
6          IntStream l1 = list.stream().mapToInt(Integer::intValue);
7          OptionalDouble out1 = l1.average();
8          System.out.println("average:" + out1.getAsDouble());
9          IntStream l2 = list.stream().mapToInt(Integer::intValue);
10         int out2 = l2.sum();
11         System.out.println("sum:" + out2);
12         IntStream l3 = list.stream().mapToInt(Integer::intValue);
13         long out3 = l3.count();
14         System.out.println("count:" + out3);
15         IntStream l4 = list.stream().mapToInt(Integer::intValue);
16         OptionalInt out4 = l4.max();
17         System.out.println("max:" + out4.getAsInt());
18         //对象 Stream(对象流)
19         System.out.println("------------------");
20         System.out.println("对象流操作:");
21         Stream<Integer> s5 = list.stream();
22         long out5 = s5.count();
23         System.out.println("count:" + out5);
24         Stream<Integer> s6 = list.stream();
25         Optional<Integer> out6 = s6.max(Integer::compareTo);
26         System.out.println("max 对象流:" + out6.get());
27     }
28 }
```

运行结果如图 7-3-7 所示。

```
原生流操作:
average:94.875
sum:759
count:8
max:201
------------------
对象流操作:
count:8
max对象流:201
```

图 7-3-7　运行结果

【说明】

程序中每完成一个最终操作都必须重新生成 Stream。第 6、9、12、15 行生成对象 Stream 并转换为原生 Stream，第 7、10、13、16 行分别调用 average( )、sum( )、count( )、max( )方法求出 Stream 元素的平均值、累加和、数量、最大值。第 8、11、14、17 行获取返回值并输出。第 21 行生成对象 Stream。第 25 行调用带参 max(Integer:: compareTo) 方法求出最大值。第 26 行调用 get( )方法获取返回值并输出。

### 2. match( )方法

match( )方法用于匹配 Stream 中的元素是否满足规则，需要提供 Predicate <T>类型参数，返回值类型为 boolean。常见的 match( )方法有 anyMatch( )、allMatch( )、noneMatch( )。

(1) anyMatch( )方法：若 Stream 中存在任意元素满足规则，则返回 true，否则返回 false。

(2) allMatch( )方法：若 Stream 中所有元素都满足规则，则返回 true，否则返回 false。

(3) noneMatch( )方法：若 Stream 中所有元素都不满足规则，则返回 true，否则返回 false。

**【例 7-3-8】** 代码如下：

```
1  public class P7_3_8 {
2      public static void main(String[] args) {
3          List<Integer> list1 = Arrays.asList(123,32,98,15,44,201);
4          boolean out1 = list1.stream().allMatch(k -> k > 100);
5          List<Integer> list2 = Arrays.asList(123,32,98,15,44,201);
6          boolean out2 = list2.stream().anyMatch(k -> k > 100);
7          List<Integer> list3 = Arrays.asList(123,32,98,15,44,201);
8          boolean out3 = list3.stream().noneMatch(k -> k > 100);
9          System.out.println("out1 = " + out1 + ",out2 = " + out2 + ",out3 = " + out3);
10     }
11 }
```

运行结果如图 7-3-8 所示。

```
out1=false,out2=true,out3=false
```

图 7-3-8　运行结果

### 3. collect( )方法

collect( )方法用于将 Stream 中的所有元素收集到可变容器中，其参数为 Collector 接口的实例。Collectors 是 Collector 接口的实现类，通常采用 Collectors 类的静态方法作为 collect( )方法的参数。常用的 Collectors 类的静态方法如下。

(1) Collectors. toSet( )：将 Stream 元素收集到集合中。

(2) Collectors. toList( )：将 Stream 元素收集到列表中。

(3) Collectors. toMap (Function <T, K>, Function <T, U>)：将 Stream 元素收集到 Map 接口中，Stream 中的每个元素均接收一个从 T 到 K 的键抽取函数及从 T 到 U 的值抽取函数。

（4）Collectors. joining( )：将字符串 Stream 中的元素拼接为字符串，通过 joining( )方法的参数指定元素之间的分隔符。

【例 7 – 3 – 9】代码如下：

```
1  public class P7_3_9 {
2      public static void main(String[] args) {
3          List<String> list1 = Arrays.asList("hello","hi","good","ok");
4          Set<String> out1 = list1.stream().filter(t -> t.startsWith("h"))
                                               .collect(Collectors.toSet());
5          System.out.println("toSet:" + out1);
6          List<String> list2 = Arrays.asList("hello","hi","good","ok");
7          String out2 = list2.stream().collect(Collectors.joining(","));
8          System.out.println("joining:" + out2);
9          List<String> list3 = Arrays.asList("hello","hi","good","ok",
                                              "ok","hi");
10         Map<String,Integer> out4 = list3.stream().distinct().collect(
          Collectors.toMap(t -> t,t -> Collections.frequency(list3, t)));
11        System.out.println("toMap:" + out4);
12     }
13 }
```

运行结果如图 7 – 3 – 9 所示。

```
toSet:[hi, hello]
joining:hello,hi,good,ok
toMap:{hi=2, hello=1, ok=2, good=1}
```

图 7 – 3 – 9　运行结果

【说明】

程序第 4 行生成 Stream，过滤出以字母“h”开头的 Stream 元素，最后将 Stream 元素收集到集合中。第 7 行生成 Stream，并以字符“,”为间隔符连接所有 Stream 元素，并返回结果字符串。第 10 行生成 Stream，并去除 Stream 中的重复元素，将 Stream 元素收集到 Map 接口中，其中 Map 键为 Stream 元素，值为 Stream 元素在 Stream 中出现的次数。

**4. reduce( )方法**

reduce( )方法用于对 Stream 中的元素通过指定方法进行归并处理，参数为 BinaryOperator<T>类型，返回值为 Optional 类型。reduce( )方法将二元函数从前两个元素开始按照指定方法进行运算，并反复将前面的运算结果与 Stream 中剩余的元素进行同样的运算。在实际应用中，求累加和、求累乘积、求最大值、集合并集等都可以使用 reduce( )方法实现。

【例 7 – 3 – 10】代码如下：

```
1  public class P7_3_10 {
2      public static void main(String[] args) {
3          List<Integer> list1 = Arrays.asList(1,2,3,4,5);
4          int out1 = list1.stream().reduce((x,y) -> x + y).get();
5          System.out.println(out1);
6      }
7  }
```

【说明】

程序第 4 行生成 Stream，调用 reduce( )方法指定 Stream 元素按照求和的方式进行运算，累加和以 Optional 类型返回，通过 get( )方法获取返回值。

**案例**

在学生成绩管理系统中，如何通过 Stream 实现课程成绩的统计？

【分析】

（1）学生成绩信息的组织。

定义封装类 StuScore，代码如下：

```
public class StuScore {
    private String id ,name,sex;
    private double chinese, math, english;
}
```

（2）使用 Map 接口组织所有学生成绩信息。

（3）录入学生成绩信息，并存入 Map 接口。

（4）定义课程名枚举类型 Subjects，代码如下：

```
enum Subjects{
    CHINESE ,MATH ,ENGLISH ;
}
```

（5）统计课程成绩。

利用 Stream 的 max( )方法、min( )方法、filter( )方法、forEach( )方法实现。

**1. 窗体设计**

窗体设计如图 7－3－10 所示。

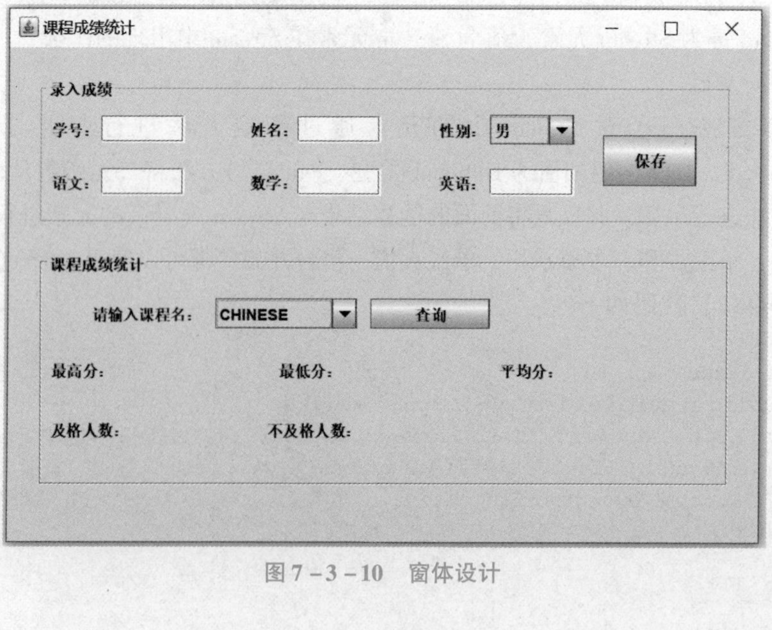

图 7－3－10　窗体设计

**2. 功能实现**

代码如下：

```
public class Example7_3 extends JFrame {
    private static final long serialVersionUID = 1L;
    private JPanel contentPane;
    private JTextField textField,textField_1,textField_3;
    private JTextField textField_4,textField_5;
    private JLabel lblNewLabel_10,lblNewLabel_11,lblNewLabel_12;
    private JLabel lblNewLabel_13,lblNewLabel_16;
    static double resultmax,resultmin,resultsum;
    static long passcnt;

    public static void main(String[] args) {
        EventQueue.invokeLater(new Runnable() {
          public void run() {
              try {
                  Example6_3 frame = new Example6_3();
                  frame.setVisible(true);
              } catch (Exception e) {
                  e.printStackTrace();
              }
          }
        });
    }
    public Example7_3() {
      setTitle("课程成绩统计");
      ……(略)
      Map<String,StuScore> scoreMap = new HashMap<String,StuScore>();

      JButton btnNewButton = new JButton("保存");
      btnNewButton.addActionListener(new ActionListener() {
          public void actionPerformed(ActionEvent e) {
              StuScore ss1 = new StuScore();
              ss1.setId(textField.getText());
              ss1.setName(textField_1.getText());
              ss1.setSex(comboBox.getSelectedItem().toString());
              ss1.setChinese(Double.parseDouble(textField_3.getText()));
              ss1.setMath(Integer.parseInt(textField_4.getText()));
              ss1.setEnglish(Integer.parseInt(textField_5.getText()));
              scoreMap.put(textField.getText(), ss1);
              textField.setText("");
              textField_1.setText("");
              textField_3.setText("");
              textField_4.setText("");
              textField_5.setText("");

          }
        });
```

```
            btnNewButton.setBounds(441, 43, 73, 39);
            panel.add(btnNewButton);
        ……(略)
            JButton btnNewButton_1 = new JButton("查询");
            btnNewButton_1.addActionListener(new ActionListener() {
                public void actionPerformed(ActionEvent e) {
                    Collection<StuScore> scores = scoreMap.values();
                    Subjects subj = (Subjects)comboBox_1.getSelectedItem();
                    switch (subj) {
                        case CHINESE :resultmax = scores.stream().max((e1,e2) ->
                          Double.compare(e1.getChinese(),e2.getChinese()))
                          .get().getChinese();
                          resultmin = scores.stream().min((e1,e2) ->
                          Double.compare(e1.getChinese(),e2.getChinese()))
                          .get().getChinese();
                          scores.stream().forEach(k ->{resultsum + =
                          k.getChinese();});
                          passcnt = scores.stream().filter(k ->k.getChinese()
                           > =60).count();
                          break;
                       case MATH :resultmax = scores.stream().max((e1,e2) ->
                          Double.compare(e1.getMath(), e2.getMath()))
                          .get().getMath();
                          resultmin = scores.stream().min((e1,e2) ->
                          Double.compare(e1.getMath(), e2.getMath()))
                          .get().getMath();
                          scores.stream().forEach(k ->{resultsum + =
                          k.getMath();});
                          passcnt = scores.stream().filter(k ->k.getMath()
                           > =60).count();
                          break;
                       default :resultmax = scores.stream().max((e1,e2) ->
                          Double.compare(e1.getEnglish(), e2.getEnglish()))
                          .get().getEnglish();   resultmin = scores.stream().min
((e1,e2) ->
                          Double.compare(e1.getEnglish(), e2.getEnglish()))
                          .get().getEnglish();
                          scores.stream().forEach(k ->{resultsum + =
                          k.getEnglish();});
                          passcnt = scores.stream().filter(k ->k.getEnglish()
                           > =60).count();
                          break;
                    }
                    lblNewLabel_10.setText("" + resultmax);
                    lblNewLabel_11.setText("" + resultmin);
                    lblNewLabel_12.setText("" + resultsum/scoreMap.size());
                    lblNewLabel_13.setText("" + passcnt);
                    lblNewLabel_16.setText("" + (scoreMap.size() - passcnt));
                }
```

```
            });
        }
    }
enum Subjects{
    CHINESE ,MATH ,ENGLISH ;
}
```

**任务工单**

<table>
<tr><td>任务名称</td><td colspan="5">图书信息管理系统——使用 Stream 方法实现图书模糊查询</td></tr>
<tr><td>班级</td><td></td><td>学号</td><td></td><td>姓名</td><td></td></tr>
<tr><td>任务要求</td><td colspan="5">图书模糊查询<br>查询关键字： java 模糊查询<br>Book [bookID=001, bookName=Java程序设计, writer=王强, press=清华大学出版社, price=50.0]<br>Book [bookID=002, bookName=Java案例教程, writer=王新, press=北京大学出版社, price=58.0]<br>Book [bookID=003, bookName=Java程序设计从入门到精通, writer=周北平, press=清华大学出版社, price=46.0]<br><br>（1）定义图书集合，代码如下：<br><br>List <Book> books = new ArrayList <Book>();<br>books.add(new Book("001","Java 程序设计","王强","清华大学出版社",50,"在库"));<br>books.add(new Book("002","Java 案例教程","王新","北京大学出版社",58,"在库"));<br>books.add(new Book("003","Java 程序设计从入门到精通","周北平","清华大学出版社",46,"在库"));<br>books.add(new Book("004","C 语言程序设计","何贝","高等教育出版社",38,"在库"));<br>books.add(new Book("005","C 语言案例教程","周华金","高等教育出版社",40,"在库"));<br>books.add(new Book("006","Python 程序设计","林琳","北京大学出版社",50,"在库"));<br><br>（2）使用 Stream 的 filter( )方法实现图书信息的模糊查询。</td></tr>
<tr><td>任务实现</td><td colspan="5">(提交模糊查询交互代码)</td></tr>
<tr><td>异常及<br>待解决<br>问题记录</td><td colspan="5"></td></tr>
</table>

## 练习题

### 1. 选择题

（1）引用 Integer 类的 parseInt( )方法，正确的表达式为（　　）。

A. Integer. parseInt　　B. Integer:: parseInt

C. Integer: parseInt　　D. ( ) -> Integer. parseInt (x)

（2）设有接口：

```
Interface sub{
      Public int subfun(int x,int y);
}
```

针对该接口的 Lambda 表达式错误的是（　　）。

A. (int x,int y) -> (return x - y;)

B. (a,b) -> {return a - b;}

C. (x,y) -> return a - b;

D. (a,b) -> a - b;

（3）以下函数接口中，含有名为 test 的方法是（　　）。

A. Predicate <T>　　B. Consumer <T>

C. Function <T, R>　　D. Supplier <T>

（4）以下方法中，属于 Stream 的最终操作的是（　　）。

A. distinct( )　　B. map( )　　C. filter( )　　D. forEach( )

### 2. 填空题

（1）

```
Stream.of(1,3,4,0,2,4).takeWhile(x->x>0).forEach(System.out::print);
```

运行结果：______

（2）

```
Stream.of(1,2,4).map(x->x*x).forEach(System.out::print);
```

运行结果：______

（3）

```
List<Integer> list = Arrays.asList(5,1,2,3);
var r = list.stream().reduce((x,y)->x+y).get();
System.out.println(r);
```

运行结果：______

（4）

```
List<String> my = Arrays.asList("buy","book","pan","car");
var x = my.stream().filter((s)->s.contains("b")).count();
System.out.println(x);
```

运行结果：______

# 第8章

# 异常处理

——学无止境，求知若渴

## 8.1 异常概述

**学习目标**

（1）理解异常的概念。
（2）了解常见的异常。

**相关知识**

### 8.1.1 异常的概念

程序在执行过程中，总会出现一些不可预期的情况，例如除法运算时除数为 0、数组下标越界、网络中断、加载的类不存在等。这些非正常的情况会导致 Java 虚拟机非正常停止，称为异常。下面通过案例认识异常。

**【例 8-1-1】**代码如下：

```
2  public class P8_1_1{
3      public static void main(String[] args){
4          int result = div(10, 0); //调用 div()方法
5          System. out .println(result);
6      }
7
8      //两个整数相除方法 div()
9      public static int div (int x, int y) {
10         int result = x /y;
11         return result;
12     }
13 }
```

运行结果如图 8-1-1 所示。

```
Exception in thread "main" java.lang.ArithmeticException: / by zero
        at tt.P8_1_1.div(P8_1_1.java:10)
        at tt.P8_1_1.main(P8_1_1.java:4)
```

图 8-1-1 运行结果

【说明】

程序发生了算数异常（ArithmeticException），第 3 行调用 div( )方法，传入了参数 0，导致运算时出现了除零异常。出现异常后，程序停止运行，Java 异常处理机制抛出异常。

### 8.1.2 异常的分类

Java 以 java. lang. Throwable 类作为异常处理的父类，Throwable 类有两个直接子类——Error 类和 Exception 类，其中 Error 代表程序中产生的错误，Exception 代表程序中产生的异

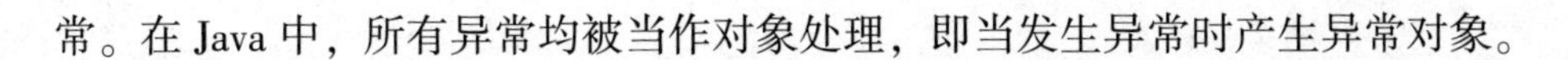

常。在 Java 中，所有异常均被当作对象处理，即当发生异常时产生异常对象。

## 一、Error 类

Error 类称为错误类，是一类比较严重的错误，无法通过修改程序本身来恢复运行，这类错误往往由硬件故障或者系统内部的错误引起。例如，汽车发动机故障导致汽车无法正常行驶，这种故障是无法通过简单的修复排除的。

## 二、Exception 类

Exception 类称为异常类，是一类通过修改程序本身可以处理的错误。Java 程序中的异常处理针对的就是该类及其子类。这类异常往往出现在程序编译或运行过程中，可以通过 Java 提供的异常处理机制检查出现的异常，并进行处理，保证程序正常运行。例如，导致汽车无法正常行驶的原因是燃油不足，这时可以通过添加燃油来解决问题。

### 8.1.3　常见的异常

Java 程序在编译过程中会出现多种多样的异常，它们大多属于 Exception 类。总的来说，异常分为非检查性异常和检查性异常两大类。

## 一、非检查性异常

非检查性异常如表 8－1－1 所示。

表 8－1－1　非检查性异常

| 异常 | 功能描述 |
| --- | --- |
| ArithmeticException | 当出现异常的运算条件时抛出的异常。例如，除零运算抛出此类的一个实例 |
| ArrayIndexOutOfBoundsException | 当用非法索引访问数组时抛出的异常。如果索引超出范围，则该索引为非法索引 |
| ArrayStoreException | 当错误类型的对象存储到一个对象数组中时抛出的异常 |
| ClassCastException | 当对象类型强制转换为不是实例的子类时抛出的异常 |
| IllegalArgumentException | 当向方法传递不合法的参数时抛出的异常 |
| NullPointerException | 当程序调用未被初始化的对象时抛出的异常 |
| NumberFormatException | 当将字符串转换成一种数值类型，但该字符串不能转换为适当格式时抛出的异常 |
| StringIndexOutOfBoundsException | 此异常由 String 方法抛出，指示索引或者为负，或者超出字符串的大小 |

## 二、检查性异常

检查性异常如表 8－1－2 所示。

表 8－1－2　检查性异常

| 异常 | 功能描述 |
| --- | --- |
| ClassNotFoundException | 当应用程序试图加载类，但找不到相应的类时抛出的异常 |
| CloneNotSupportedException | 当调用 Object 类中的 clone( ) 方法克隆对象，但该对象的类无法实现 Cloneable 接口时抛出的异常 |
| IllegalAccessException | 当拒绝访问一个类时抛出的异常 |
| InstantiationException | 当接口或抽象类无法被实例化时抛出的异常 |
| InterruptedException | 当一个线程被另一个线程中断时抛出的异常 |
| NoSuchFieldException | 当请求的变量不存在时抛出的异常 |
| NoSuchMethodException | 当请求的方法不存在时抛出的异常 |

# 8.2 异常的处理方法

## 学习目标

（1）掌握 try…catch…finally 模块的异常捕获方法。

（2）掌握 throws 抛出异常。

（3）掌握 throw 自定义异常。

## 相关知识

### 8.2.1 捕获异常

为了保证程序有效地运行，需要对发生的异常进行相应的处理。当某个方法抛出异常时，有两种处理方式：一种是在当前方法中进行捕获，并处理该异常；另一种是将异常向上抛出，由方法调用者处理该异常。

#### 一、try…catch 模块

**1. 捕获单个异常**

try…catch 模块用于捕获异常，可能发生异常的语句存放在 try 代码块中，catch 代码块在 try 代码块之后，用于捕获异常并处理异常的语句写在 catch 代码块中。try…catch 模块的语法格式如下：

```
try
{
   //程序代码
}catch(ExceptionName1 e1){
   //catch 代码块
}
```

**【例 8-2-1】** 使用 try…catch 模块解决问题。代码如下：

```
1  public class P8_2_1{
2      public static void main(String[] args){
3          try{
4              int result = div(10,0); //调用 div()方法
5              System.out.println(result);
6          } catch(Exception e){ //对异常进行处理
7              System.out.println("异常:" + e.getMessage());
8          }
9          System.out.println("程序继续向下执行...");
```

```
10      }
11      //两个整数相除的方法
12      public static int div (int x, int y) {
13           int result = x /y;
14           return result;
15           }
16  }
```

运行结果如图 8-2-1 所示。

```
捕获的异常信息为：/ by zero
程序继续向下执行...
```

图 8-2-1　运行结果

【说明】

从运行结果可以看出，catch 代码块对 try 代码块中可能发生的异常进行了处理。在 try 代码块中发生“除零异常”时，程序会转而执行 catch 代码块中的代码，此时控制台输出 Exception 对象的 getMessage( )方法返回的异常信息“/by zero”。在 catch 代码块对异常处理完毕后，主程序仍会向下执行，不会因为出现异常而终止执行。

**小贴士**

在 try 代码块中，发生异常语句后面的代码是不会被执行的，例如【例 8-2-1】的程序中第 7 行代码的打印语句就没有执行。

### 2. 捕获多种异常

当代码块中出现多种异常需要捕获时，可以使用多个 catch 代码块分别捕获这些异常，也可以使用父类异常捕获所有异常。值得注意的是，当同时存在子类异常捕获和父类异常捕获时，父类异常捕获必须放在子类异常捕获之后，否则子类异常将被父类异常捕获。捕获多种异常的语法格式如下：

```
try
{
//程序代码
}catch(ExceptionName1 e1){
//catch 块
}catch(ExceptionName2 e2){
//catch 块
}
```

或

```
try
{
//程序代码
```

```
}catch(Exception e1){
//catch块

}
```

【例8-2-2】代码如下：

```
1  public class P8_2_2 {
2      int div(int x,int y) {
3          int result = x/y;
4          return result;
5      }
6      public static void main(String[] args) {
7          P8_2_2 d6 = new P8_2_2 ();
8          int x = 0;
9          try {
10             x = d6.div(4, 1);
11             int array[] = new int [4];
12             System. out .println(array[4]);
13         } catch (ArithmeticException e) {
14             System. out .println("异常原因:" + e.getMessage());
15         } catch (ArrayIndexOutOfBoundsException e1){
16             System. out .println("异常原因:数组越界了");
17         }
18         System. out .println("x = " + x);
19     }
20 }
```

【说明】

程序第13行捕获了算术异常，第15行捕获了数组下标越界异常。同时捕获两种异常也可以通过父类异常来实现，代码如下：

```
6      public static void main(String[] args) {
7          P8_2_2 d6 = new P8_2_2 ();
8          int x = 0;
9          try {
10             x = d6.div(4, 1);
11             int array[] = new int [4];
12             System. out .println(array[4]);
13         } catch (Exception e) {
14             System. out .println("异常原因:" + e.getMessage());
15         }
16         System. out .println("x = " + x);
17     }
```

【说明】

当异常捕获类型同时存在子类异常捕获和父类异常捕获时，父类异常必须放在所有异常捕获的最后面，代码如下：

```
6       public static void main(String[] args) {
7           P8_2_2 d6 = new P8_2_2 ();
8           int x = 0;
9           try {
10              x = d6.div(4, 1);
11              int array[] = new int [4];
12              System. out .println(array[4]);
13          } catch (ArithmeticException e) {
14              System. out .println("异常原因:" + e.getMessage());
15          } catch (Exception e1) {
16              System. out .println("异常原因:" + e.getMessage());
17          }
18          System. out .println("x = " + x);
19      }
```

## 二、try…catch…finally 模块

finally 关键字用来创建在 try 代码块或 catch 代码块后面执行的代码块。无论异常是否发生，finally 代码块中的代码总会被执行。因此，finally 代码块通常用于执行一些清理类型及收尾善后性质的语句，例如文件存取后的 IO 流的关闭。try…catch…finally 模块的语法格式如下：

```
try{
        …//执行代码块
    }catch(ExceptionType1 e1){
        …//对异常类型 1 的处理
    }catch(ExceptionType2 e2){
        …//对异常类型 2 的处理
    }
    …
    finally{
        …
    }
```

以【例 8-2-1】的错误为例，使用 try…catch…finally 模块解决问题，代码如下：

```
1  public class P8_2_1{
2      public static void main(String[] args){
3          try {
4              int result  = div (10, 0); //调用 div()方法
5              System. out .println(result);
6          } catch (Exception e) { //对捕获到的异常进行处理
7              System. out .println("捕获的异常信息为:" + e.getMessage());
8              return ; //结束 catch 代码块语句
9          } finally {
10             System. out .println("进入 finally 代码块");
```

```
11          }
12          System.out.println("程序继续向下执行…");
13          }
14          public static int div (int x, int y) {//两个整数相除的方法
15              int result = x /y; //定义一个变量 result 记录两个数相除的结果
16              return result; //将结果返回
17              }
18  }
```

运行结果如图 8-2-2 所示。

```
捕获的异常信息为：/ by zero
进入finally代码块
```

图 8-2-2　运行结果

【说明】

在程序中，catch 代码块中增加了一个 return 语句，用于结束 catch 代码块的执行，此时第 12 行代码将不被执行，而 finally 代码块中的代码仍会被执行，不受 return 语句的影响。也就是说，无论程序是发生异常还是使用 return 语句结束，finally 代码块中的代码都会被执行。

### 8.2.2　抛出异常

若出现异常的方法不处理异常的情况，则可以在方法中使用 throws 关键字将异常抛出，交由调用者处理，允许一次性抛出多个异常，异常间使用逗号分隔。抛出异常的语法格式如下：

```
访问控制修饰符 返回值数据类型 方法名(形参列表) throws 异常类型(异常类){
  //实际书写的位置是形参列表的最后一个括号后面
}
```

以【例 8-2-1】中的错误为例，使用 throws 关键字解决问题，代码如下：

```
1  package P8;
2  public class P8_2_1{
3      public static void main(String[] args){
4          int result = div (10, 0); //调用 div ()方法
5          System.out.println(result);
6      }
7      //两个整数相除的方法
8      public static int div(int x, int y) throws ArithmeticException {
9          int result = x /y;
10         return result;
11     }
12 }
```

运行结果如图 8-2-3 所示。

```
Exception in thread "main" java.lang.ArithmeticException: / by zero
	at tt.P8_2_1.div(P8_2_1.java:9)
	at tt.P8_2_1.main(P8_2_1.java:4)
```

图 8-2-3　运行结果

【说明】

div()方法没有对除零异常进行处理，而是选择抛出一个算术异常，但调用 div()方法的主函数并没有对算术异常进行处理，这就导致程序无法正常运行。

### 8.2.3　自定义异常类

在程序开发中，除了可以捕获 Java 内置的异常类外，还可以捕获用户自定义的异常类。用户自定义的异常类只需要继承 Exception 类即可。使用自定义异常类，分为以下两个步骤。

（1）创建自定义异常类。

（2）在方法中通过 throw 关键字抛出自定义异常类对象。

使用自定义异常类的语法格式如下：

```
访问控制修饰符 返回值数据类型 方法名(形参列表) {
    throw new Exception("异常信息");
}
```

以【例 8-2-1】中的错误为例，使用 throw 关键字解决问题，代码如下：

```
1  public class P8_2_1{
2      public static void main(String[] args){
3          int result = div(10, 0);
4          System. out .println(result);
5      }
6      public static int div(int x, int y) {//两个整数相除的方法
7          try {
8              if (y ==0){
9                  throw new Exception("除零异常!");
10             }
11             int result = x /y;
12             return result;
13         }catch (Exception e) { //对异常进行处理
14             System. out .println("捕获的异常信息为:" + e.getMessage());
15         }
16         return y;
17         }
18 }
```

运行结果如图 8-2-4 所示。

```
捕获的异常信息为：除零异常！
0
```

图 8-2-4　运行结果

【说明】

在上述代码中，在 div()方法中事先声明了当 y 等于 0 时将抛出自定义异常类对象，以避免在主函数调用时出现异常，导致程序无法运行。

**案例**

在各类管理系统中不可避免地要进行运算，用户由于误操作输入非法数据会导致运算无法正常进行，这种异常如何处理？

**1. 异常分析**

(1) 用户输入非法数据（图 8-2-5）。

(2) 除数为 0（图 8-2-6）。

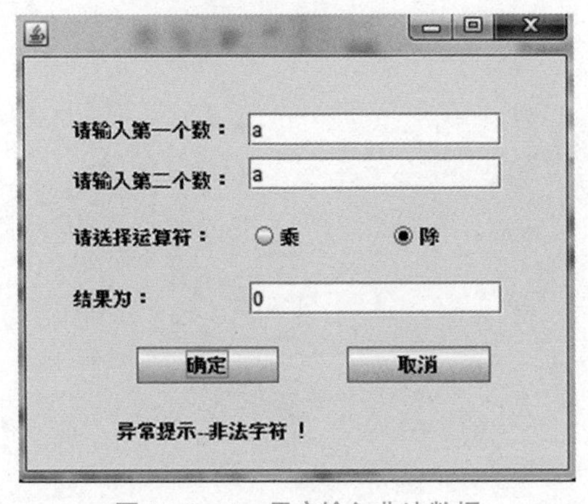


图 8-2-5　用户输入非法数据

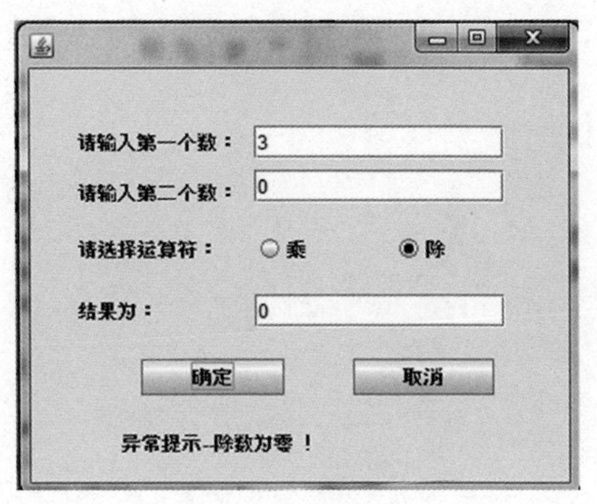


图 8-2-6　除数为 0

**2. 功能实现**

代码如下：

```
public class ExceptionEx extends JFrame {
    private JPanel contentPane;
    private JTextField textField , textField_1,textField_2;
    JLabel label_4;
    ……
    public ExceptionEx() {
        ……(略)
        JButton button = new JButton("确定");
        button.addActionListener(new ActionListener() {
            public void actionPerformed(ActionEvent arg0) {
                int result =0;
                try {
                    String s1 =textField.getText();
                    String s2 =textField_1.getText();
                    int n1 = Integer.parseInt(s1);
                    int n2 = Integer.parseInt(s2);
                    if (radioButton.isSelected())
                        result =n1 * n2;
```

```
                if(radioButton_1.isSelected())
                    result = n1/n2;
            } catch(NumberFormatException e) {
                label_4.setText(label_4.getText() + " --非法字符");
            } catch(ArithmeticException e1){
                label_4.setText(label_4.getText() + " --除数为零");
            }
            textField_2.setText("" + result);
        }
    });
    ……(略)
}}
```

任务工单

| 任务名称 | 图书信息管理系统——图书、杂志业务处理 | | | | |
|---|---|---|---|---|---|
| 班级 | | 学号 | | 姓名 | |

任务要求

（1）定义图书集合，查询指定图书编号信息，代码如下：

```
List<Book> books = new ArrayList<Book>();
books.add(new Book("001","Java 程序设计","王强","清华大学出版社",
50,"在库"));
books.add(new Book("002","Java 案例教程","王新","北京大学出版社",
58,"在库"));
books.add(new Book("003","Java 程序设计从入门到精通","周北平","清
华大学出版社",46,"在库"));
books.add(new Book("004","C 语言程序设计","何贝","高等教育出版社",
38,"在库"));
books.add(new Book("005","C 语言案例教程","周华金","高等教育出版
社",40,"已借"));
books.add(new Book("006","Python 程序设计","林琳","北京大学出版
社",50,"已借"));
```

（2）当完成更新时，输入的单价若非数值型，则捕获 NumberFormatException 异常，并给出提示，如下图所示。

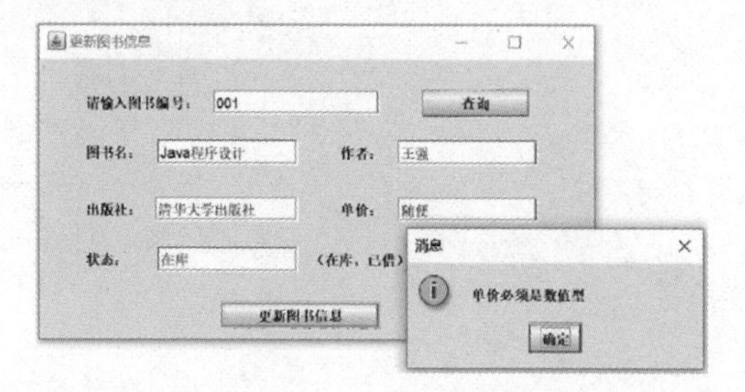


（3）当完成更新时，输入的状态若非“在库”或“已借”，则捕获自定义异常 throw new Exception("图书状态必须是\"在库\"或者\"已借\"")，并给出提示，如下图所示。

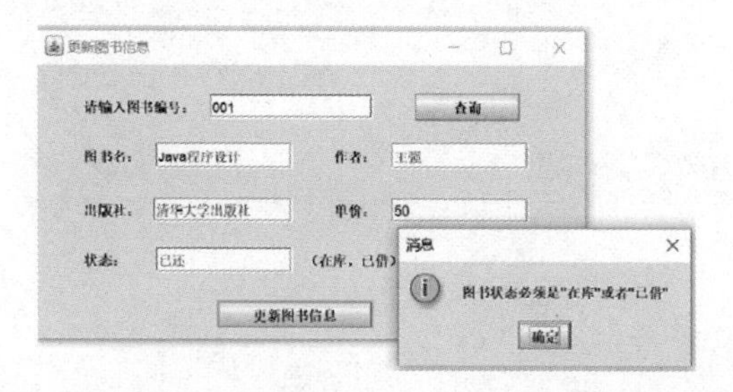

**续表**

<table>
<tr><td>任务名称</td><td colspan="5">图书信息管理系统——图书、杂志业务处理</td></tr>
<tr><td>班级</td><td></td><td>学号</td><td></td><td>姓名</td><td></td></tr>
<tr><td>任务实现</td><td colspan="5">（1）实现查询功能，代码如下：<br><pre>JButton btnNewButton = new JButton("查询");
    btnNewButton.addActionListener(new ActionListener() {
        public void actionPerformed(ActionEvent e) {



        }
    }</pre><br>（2）实现更新功能并实现异常处理，代码如下：<br><pre>JButton btnNewButton_1 = new JButton("更新图书信息");
    btnNewButton_1.addActionListener(new ActionListener() {
        public void actionPerformed(ActionEvent e) {



        }
    }</pre></td></tr>
<tr><td>异常及待解决问题记录</td><td colspan="5"></td></tr>
</table>

## 练习题

### 1. 选择题

（1）异常包含下列哪些内容？（　　）

A. 程序运行过程中遇到的事先没有预料到的情况

B. 程序中的语法错误

C. 程序中的编译错误

D. 以上都是

（2）Java 中用来抛出异常的关键字是（　　）。

A. try　　B. catch　　C. throw　　D. finally

（3）关于异常，下列说法正确的是（　　）。

A. 异常是一种对象

B. 一旦程序运行，异常就将被创建

C. 为了保证程序的运行速度，要尽量避免异常控制

D. 以上说法都不对

（4）（　　）类是所有异常类的父类。

A. Throwable　　B. Error　　C. Exception　　D. AWTError

（5）在 Java 中，（　　）是异常处理的出口。

A. try 代码块　　B. catch 代码块　　C. finally 代码块　　D. 以上说法都不对

（6）对于 catch 代码块的排列，下列说法正确的是（　　）。

A. 父类在先，子类在后

B. 子类在先，父类在后

C. 有继承关系的异常不能在同一个 try 代码块内

D. 先有子类，其他类的顺序无关紧要

（7）在异常处理中，释放资源、关闭文件、关闭数据库等由（　　）完成。

A. try 代码块　　B. catch 代码块　　C. finally 代码块　　D. throw 代码块

（8）当方法遇到异常又不知如何处理时，会（　　）。

A. 捕获异常　　B. 抛出异常　　C. 声明异常　　D. 嵌套异常

### 2. 填空题

（1）在 Java 中，程序运行时出现的意外事件称为__________，处理异常的过程称为__________。

（2）所有异常类都直接或间接地继承__________类。

（3）运行时异常表示程序运行时发现的由__________抛出的各种异常，这些异常对应__________。

（4）catch 代码块都带一个参数，该参数是某个异常的类及其变量名，catch 代码块用该参数与__________进行匹配。

（5）按照处理方式的不同，可以将异常分运行异常、捕获异常、__________和声明

异常几种。

（6）对于编程语言而言，错误一般有编译错误和________错误两类。

（7）Java 中的所有异常都继承 Throwable 类。Throwable 类有两个直接子类：Exception 类和________________类。

（8）异常类对象从产生到被传递给 Java 运行系统的过程称为______________异常。

**3. 编程题**

（1）对下述代码进行异常处理。

```
public  class   ExceptionDemo{
     public static void main(String[ ]args){
        int   a =2,b =0;
        System.out.printin("This is an exception." + a/b);
          System.out.println("Finished");
        }
}
```

（2）用户自定义异常类练习。

①创建 AgeException 异常类，继承 Exception 类。

②创建 Example 类，在 Example 类中创建一个 readAge( )方法，该方法用于从键盘获取年龄，检查年龄是否为 18 ~22 岁，若超出该范围，则抛出 AgeException 异常。

③在主程序中捕获自定义异常类对象，并进行相应处理。

第9章

# IO流

——勤学若渴，不负韶华

# 9.1 字节流

## 学习目标

（1）了解流的概念。

（2）掌握 InputStream 类、OutputStream 类的使用方法。

（3）掌握使用 FileInputStream 类、FileOutputStream 类操作文件的方法。

## 相关知识

### 9.1.1 流的概念

#### 一、IO 流

IO 流基于流的概念，将数据的输入和输出看作一个连续的流。IO 流提供了一条通道，可以使用这条通道把源中的字节序列送到目标。Java 以数据流的形式处理输入和输出，程序从指向源的输入流中读取源中的数据，源可以是文件、网络连接、压缩包或者其他数据源；程序通过向输出流中写入数据把信息传递到目标，输出流的目标可以是文件、网络连接、压缩包、控制台或者其他数据输出目标。

IO 流示意如图 9－1－1 所示。

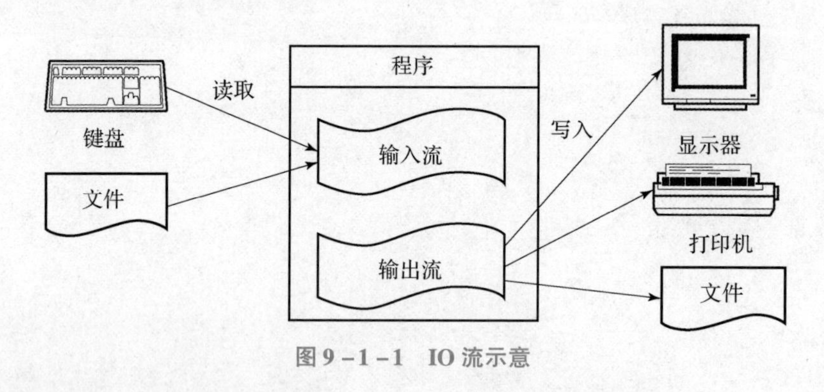


图 9－1－1　IO 流示意

#### 二、IO 流的分类

Java 中的 IO 流可以根据数据类型和流的方向进行分类。

**1. 按数据类型分类**

（1）字节流（Byte Stream）：以字节为单位读写数据，适用于处理二进制数据，如图像、音频、视频等。常见的字节流类有 InputStream 和 OutputStream。

（2）字符流（Character Stream）：以字符为单位读写数据，适用于处理文本数据。字符

流会自动进行字符编码和解码，可以处理多国语言字符。常见的字符流类有 Reader 和 Writer。

**2. 按流的方向分类**

(1) 输入流（Input Stream）：用于读取数据。输入流从数据源读取数据，如文件、网络连接等。常见的输入流类有 FileInputStream、ByteArrayInputStream、SocketInputStream 等。

(2) 输出流（Output Stream）：用于写入数据。输出流将数据写入目标，如文件、数据库、网络连接等。常见的输出流类有 FileOutputStream、ByteArrayOutputStream、SocketOutputStream 等。

### 9.1.2 文件类

## 一、文件类的概念

IO 流可以对文件的内容进行读写操作，在应用程序中经常对文件本身进行一些常规操作，例如创建一个文件、判断硬盘中的某个文件是否存在等。针对文件的这类操作，JDK 提供了一个 File 类，该类封装了一个路径，这个路径可以是从系统盘符开始的绝对路径，例如“D:\file\a. txt”，也可以是相对于当前目录而言的相对路径，例如“src\Hello. java”。File 类内部封装的路径可以指向一个文件，也可以指向一个目录，File 类提供了针对这些文件或目录的一些常规操作方法。表 9－1－1 所示为 File 类的常用构造方法。

表 9－1－1　File 类的常用构造方法

| 方法 | 功能描述 |
|---|---|
| File(String pathname) | 通过指定的一个字符串型的文件路径创建一个新的 File 类对象 |
| File(String parent，String child) | 根据指定的一个字符串型的父路径和一个字符串型的子路径（包括文件名称）创建一个 File 类对象 |
| File(File parent，String child) | 根据指定的 File 类的父路径和字符串型的子路径（包括文件名称）创建 一个 File 类对象 |

## 二、文件类的常用方法

File 类提供了一系列方法，用于操作其内部封装的路径指向的文件或者目录，例如判断文件/目录是否存在，创建、删除文件/目录等。表 9－1－2 所示为 File 类的常用方法。

表 9－1－2　File 类的常用方法

| 方法 | 功能描述 |
|---|---|
| boolean exists() | 判断 File 类对象对应的文件或目录是否存在，若存在则返回 true，否则返回 false |
| boolean delete() | 删除 File 类对象对应的文件或目录，若删除成功则返回 true，否则返回 false |

续表

| 方法 | 功能描述 |
| --- | --- |
| boolean createNewFile( ) | 当 File 类对象对应的文件不存在时，该方法将新建一个此 File 类对象所指定的新文件，若创建成功则返回 true，否则返回 false |
| String getNameO | 返回 File 类对象表示的文件或文件夹的名称 |
| String getPath( ) | 返回 File 类对象对应的路径 |
| String getAbsolutePath( ) | 返回 File 类对象对应的绝对路径（在 UNIX/Linux 等系统中，如果路径是以正斜线(/)开始的，则这个路径是绝对路径；在 Windows 等系统中，如果路径是从盘符开始的，则这个路径是绝对路径） |
| String getParent( ) | 返回 File 类对象对应目录的父目录（即返回的目录不包含最后一级子目录） |
| boolean canRead( ) | 判断 File 类对象对应的文件或目录是否可读，若可读则返回 true，否则返回 false |
| boolean canWriteO | 判断 File 类对象对应的文件或目录是否可写，若可写则返回 true，否则返回 false |
| boolean isFile( ) | 判断 File 类对象对应的是否是文件（不是目录），若是文件则返回 true，否则返回 false |
| boolean isDirectoryO | 判断 File 类对象对应的是否是目录（不是文件），若是目录则返回 true，否则返回 false |
| boolean isAbsolute( ) | 判断 File 类对象对应的文件或目录是否是绝对路径 |
| long lastModified( ) | 返回 1970 年 1 月 1 日 0 时 0 分 0 秒到文件最后修改时间的毫秒值 |
| long length( ) | 返回文件内容的长度 |
| String[ ] list( ) | 列出指定目录的全部内容，只列出名称 |
| File[ ] listFiles( ) | 返回一个包含 File 类对象所有子文件和子目录的 File 类数组 |

【例 9-1-1】代码如下：

```
1  import java.io.File;
2  public class P9_1_1{
3      public static void main(String[ ] args) throws Exception {
4          //创建 File 类对象,表示一个文件
```

```
5        File file = new File("example.txt");
6        //获取文件名称
7        System.out.println("文件名称:" + file.getName());
8        //获取文件的相对路径
9        System.out.println("文件的相对路径:" + file.getPath());
10       //获取文件的绝对路径
11       System.out.println("文件的绝对路径:" + file.getAbsolutePath());
12    }
13 }
```

运行结果如图 9－1－2 所示。

```
文件名称:example.txt
文件的相对路径:example.txt
文件的绝对路径:C:\Users\wgd\eclipse-workspace\Test\example.txt
```

图 9－1－2　运行结果

## 9.1.3　字节流介绍

### 一、字节流概述

**1. 字节流的概念**

在计算机中，无论是文本、图片、音频还是视频，所有文件都是以二进制（字节）形式存储的，IO 流中针对字节的输入和输出提供了一系列流，统称为字节流。JDK 提供了两个抽象类 InputStream 和 OutputStream，它们是字节流的顶级父类。为了方便理解，可以把 InputStream 类和 OutputStream 类比作两根“水管”，把 InputStream 类看作一个输入管道，把 OutputStream 类看作一个输出管道，数据通过 InputStream 类从源设备输入程序，通过 OutputStream 类从程序输出到目标设备，从而实现数据的传输。由此可见，IO 流中的输入和输出都是相对于程序而言的。

**2. 字节流的常用方法**

在 JDK 中，InputStream 类和 OutputStream 类是抽象类。因此，针对不同的功能，InputStream 类和 OutputStream 类提供了不同的子类。字节流类的继承关系如图 9－1－3 所示。

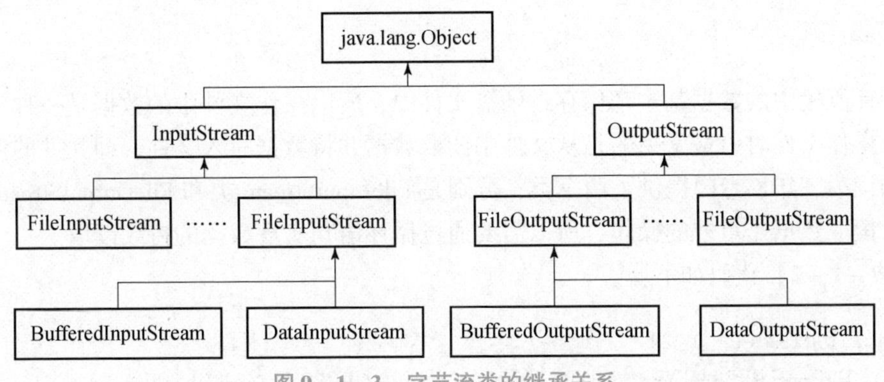


图 9－1－3　字节流类的继承关系

InputStream 类和 OutputStream 类提供了一系列与读写数据相关的方法，如表 9－1－3、表 9 －1 －4 所示。

表 9－1－3　InputStream 类的常用方法

| 方法 | 功能描述 |
| --- | --- |
| int read( ) | 从输入流读取一个 8 位的字节，把它转换为 0～255 的整数，并返回这一整数 |
| int read(byte[ ]b) | 从输入流读取若干字节，把它们保存到参数 b 指定的字节数组中，返回的整数表示读取字节的数目 |
| int read(byte[ ]b，int off，int len) | 从输入流读取若干字节，把它们保存到参数 b 指定的字节数组中，off 指定字节数组开始保存数据的起始下标，len 表示读取的字节数目 |
| void close( ) | 关闭此输入流并释放与该输入流相关的所有系统资源 |

表 9－1－4　OutputStream 类的常用方法

| 方法 | 功能描述 |
| --- | --- |
| void write(int b) | 向输出流写入一字节 |
| void write(byte[ ] b) | 将参数 b 指定的字节数组的所有字节写入输出流 |
| void write(byte[ ] b，int off，int len) | 将指定的 byte 数组中从偏移量 off 开始的 len 字节写入输出流 |
| void flush( ) | 刷新此输出流并强制写出所有缓冲的输出字节 |
| void close( ) | 关闭此输出流并释放与此输出流相关的所有系统资源 |

无论是 InputStream 类的 read( )方法还是 OutputStream 类的 write( )方法，引入字节数组作为参数，一次性读写若干字节都会提高读写效率。OutputStream 类的 flush( )方法用于将当前输出流缓冲区（通常是字节数组）中的数据强制写入目标设备，此过程称为刷新。进行 IO 流操作时，当前 IO 流会占用一定的内存，为了不造成系统资源的浪费，在 IO 流操作结束后，应该调用其 close( )方法关闭 IO 流，从而释放当前 IO 流所占用的系统资源。

## 二、字节流读写文件

由于计算机中的数据基本都保存在硬盘文件中，所以操作文件中的数据是一种很常见的操作。在操作文件时，最常见的是从文件中读取数据并将数据写入文件，即文件的读写。针对文件的读写，JDK 专门提供了两个类，分别是 FileInputStream 类和 FileOutputStream 类。由于从文件读取数据是重复的操作，所以需要通过循环语句实现数据的持续读取。

**【例 9－1－2】** 代码如下：

```
1  public class P9_1_2{
2      public static void main(String[ ] args) throws Exception {
```

```
3           FileOutputStream out  = new FileOutputStream("d:\\test.txt");
4           String str = "不积跬步,无以至千里";
5           //将字符串转换为字节数组
6           byte[] b = str.getBytes();
7           for (int i = 0; i < b.length; i ++) {
8                   out.write(b[i]);//逐个字节写入文件
9           }
10          out.close();//关闭输出流
11          System.out.println("成功写入数据!");
12          //创建一个文件字节输入流
13          FileInputStream in = new FileInputStream("d:\\test.txt");
14          int n = 0;
15          //定义一个字节数组,每次读入1024个字节数据
16          byte[] x =new byte[1024];
17          //读入字节数n不为-1,表示文件尚未读取完毕,则读入数据并保存于字节数组x中
18          while ((n = in.read(x))! = -1) {//循环判断是否读到文件尾
19              //将字节数组转为字符串,并输出
20              System.out.println(new String(x));
21          }
22          in.close();//关闭输入流
23      }
24  }
```

运行结果如图 9-1-4 所示。

```
成功写入数据!
hello java
```

图9-1-4 运行结果

【说明】

程序第 3 行创建指向 D 盘“test.txt”的字节输出流，若“test.txt”文件不存在，则系统将自动创建该文件，若“test.txt”文件已经存在，则该文件中的数据首先被清空，再写入新的数据。需要注意的是，路径字符串“\”需要转义，必须书写为“\\”，但也可以用“/”代替“\\”，即第3行写为“`FileOutputStream out = new FileOutputStream("d:/test.txt")`”。

**案例**

复制图片。

【分析】

(1) 图片文件的原始位置及目标位置：将 D 盘根目录下的“doctor.jpg”图片复制到 C 盘根目录下。

(2) 使用 FileInputStream 类、FileOutputStream 类实现。

**功能实现**

代码如下：

```
import java.io.*;
public class FileStreamCopy {
    public static void main(String[] args) {
        FileInputStream fis =null;
        FileOutputStream fos =null;
```

```
        BufferedInputStream bfinput = null ;
        BufferedOutputStream bfoutput = null ;
        File f1 = new File("d: \\doctor.jpg");
        File f2 = new File("d: \\doctorcopy0000.jpg");
        try {
            fis = new FileInputStream(f1);
            fos = new FileOutputStream(f2);
            bfinput = new BufferedInputStream(fis);
            bfoutput = new BufferedOutputStream(fos);
            int ch = 0;
            while ((ch = bfinput.read()) != -1)
                bfoutput.write(ch);
        } catch (Exception e) {
            //TODO Auto - generated catch block
            e.printStackTrace();
        }finally {
            try {
                if (bfinput != null )
                    bfinput.close();
                if (bfoutput != null )
                    bfoutput.close();
            } catch (IOException e) {
                //TODO Auto - generated catch block
                e.printStackTrace();
            }
        }
    }
}
```

## 任务工单

<table>
<tr><td>任务名称</td><td colspan="5">图书信息管理系统——图书信息保存</td></tr>
<tr><td>班级</td><td></td><td>学号</td><td></td><td>姓名</td><td></td></tr>
<tr><td>任务要求</td><td colspan="5"><br>使用字节流，将多行文本框中的图书信息保存至本机 D 盘中的“books. txt”文件（d:\books. txt）中。</td></tr>
<tr><td>任务实现</td><td colspan="5"></td></tr>
<tr><td>异常及待解决问题记录</td><td colspan="5"></td></tr>
</table>

# 9.2 字符流

## 学习目标

（1）掌握 FileReader 类、FileWriter 类的使用方法。

（2）掌握使用 FileReader 类、FileWriter 类操作文件的方法。

## 相关知识

### 9.2.1 字符流介绍

#### 一、字符流概述

**1. 字符流的概念**

尽管字节流提供了足够的方法处理任何类型的输入和输出操作，但它们不能直接操作 Unicode 字符。在实际应用中，经常会出现直接操作字符的情况，此时若以字节流方法操作字符，不但效率不高，而且容易出错，因此 JDK 提供了字符流。与字节流一样，字符流也有两个抽象的顶级父类，分别是 Reader 类和 Writer 类。其中，Reader 类为字符输入流类，用于从某个源设备读取字符；Writer 类为字符输出流类，用于向某个目标设备写入字符。

字符流的继承关系与字节流的继承关系类似，很多子类都是成对（输入流和输出流）出现的。其中，FileReader 类和 FileWriter 类用于读写文件，BufferedReader 类和 BufferedWriter 类是具有缓冲功能的字符流类，使用它们可以提高读写效率。字符流类的继承关系如图 9－2－1 所示。

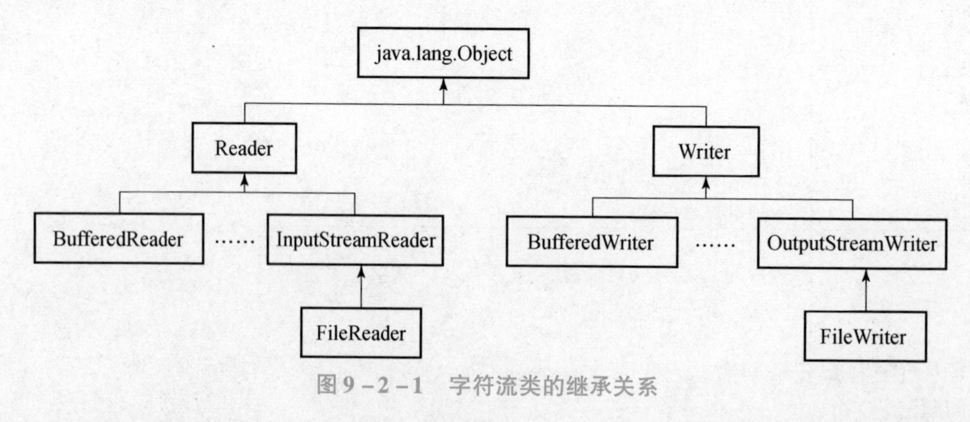


图 9－2－1 字符流类的继承关系

**2. 字符流的常用方法**

Reader 类和 Writer 类提供了一系列与读写数据相关的方法，如表 9－2－1、表 9－2－2 所示。

表 9-2-1　Reader 类的常用方法

| 方法 | 功能描述 |
|---|---|
| int read( ) | 读取单个字符，以 int 类型返回 |
| int read(char[ ] cbuf) | 读取字符放入数组 cbuf |
| int read(char[ ] cbuf,int offset,int length) | 读取字符放入数组的指定位置 |
| void close( ) | 关闭字符输入流 |
| long skip(long n) | 跳过 n 个字符 |

表 9-2-2　Writer 类的常用方法

| 方法 | 功能描述 |
|---|---|
| int write(int c) | 输出单个字符 |
| int write(char[ ] cbuf) | 输出字符数组 cbuf |
| int write(char[ ] cbuf,int offset,int length) | 输出数组中从指定位置开始的 length 个字符 |
| void close( ) | 关闭字符输出流 |
| void flush( ) | 刷新输出流并强制写出所有缓冲字符 |

## 二、字符流操作文件

在程序开发中，经常需要对文本文件的内容进行读取，如果想从文件中直接读取字符，便可以使用字符输入流类 FileReader，通过 FileReader 类可以从关联的文件中读取一个或一组字符。如果要将字符写入文件，则可以使用字符输出流 FileWriter。

**【例 9-2-1】**代码如下：

```
1  public class P9_2_1{
2      public static void main(String[ ] args) throws Exception {
3          //创建一个 FileWriter 类对象用于向文件中写入数据
4          FileWriter writer = new FileWriter("d:\\writer.txt");
5          String str = "hello java!";
6          writer.write(str); //将字符数据写入文本文件
7          writer.write("\r\n"); //将输出语句换行
8          writer.close(); //关闭字符输出流,释放资源
9          //创建一个 FileReader 类对象用于读取文件中的字符
10         FileReader reader = new FileReader("d:\\writer.txt");
11         int ch; //定义一个变量用于记录读取的字符
12         while ((ch = reader.read()) != -1) {//循环判断是否读取到文件的末尾
13             System.out.print((char) ch); //不是字符流末尾就转为字符打印
14         }
15         reader.close(); //关闭字符输入流,释放资源
16     }
17 }
```

运行结果如图 9－2－2 所示。

```
成功写入数据!
hello java
```

图 9－2－2　运行结果

### 9.2.2　转换流

在程序开发中，有时需要将字符流转换为字节流进行操作，Java 提供了转换流类 InputStreamReader 和 OutputStreamWriter 实现这个功能。

#### 一、转换输入流（InputStreamReader）

从字符流的继承关系可以发现，InputStreamReader 类是 Reader 类的子类。它的构造函数以一字节输入流作为参数，生成相应的 UTF－8 字符，其常用构造方法如下。

（1）public InputStreamReader(InputStream in)：创建转换输入流，按默认字符集的编码从输入流读取数据。

（2）public InputStreamReader(InputStream in, Charset c)：创建转换输入流，按指定字符集的编码从输入流读取数据。

为了提高读取的效率，可以使用 BufferedReader 类对字符流进行再包装（图 9－2－3），进而调用其 readLine()方法读取一行字符串。

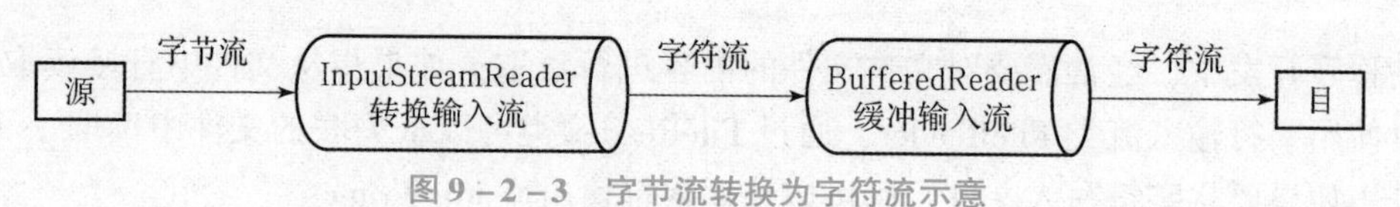


图 9－2－3　字节流转换为字符流示意

【例 9－2－2】代码如下：

```
1  import java.io.*;
2  public class P9_2_2 {
3    public static void main(String[] args) {
4      try {
5          FileInputStream fis = new FileInputStream("test.txt");
6          InputStreamReader isr = new InputStreamReader(fis,"UTF-8");
7          BufferedReader br = new BufferedReader(isr);
8          String s = "";
9          while ((s = br.readLine())! = null)
10             System.out.print(s);
11             br.close();
12     } catch (Exception e) {
13         e.printStackTrace();
14     }
15   }
16 }
```

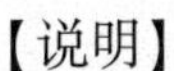

【说明】

程序第5行使用字节流方法读取工程内部文件“test. txt”，第6行以UTF－8字符编码方式将字节输入流转换为字符输入流，第7行将字符输入流转换为缓冲字符输入流，第9行循环调用缓冲字符输入流的readLine()方法读取一行数据并输出，第11行关闭缓冲字符输入流。

## 二、转换输出流（OutputStreamWriter）

从字符流的继承关系可以发现，OutputStreamWriter类是Writer类的子类。OutputStreamWriter类将UTF－8字符转换为指定字符编码形式写入字节输出流，其常用构造方法如下。

（1）public OutputStreamReader(OutputStream out)：创建转换输出流，按默认字符集的编码向输出流写入数据。

（2）public OutputStreamReader(OutputStream out, Charset c)：创建转换输出流，按指定字符集编码向输出流写入数据。

**【例9－2－3】** 代码如下：

```
1  import java.io.*;
2  public class P9_2_3 {
3      public static void main(String[] args) {
4          try {
5              FileOutputStream fos = new FileOutputStream("text.txt");
6              OutputStreamWriter osw = new OutputStreamWriter(fos,"UTF-8");
7              BufferedWriter bw = new BufferedWriter(osw);
8              bw.write("这是一个缓冲输出流");
9              bw.close();
10         } catch (Exception e) {
11             e.printStackTrace();
12         }
13     }
14 }
```

【说明】

程序第5行定义写入工程内部文件“test. txt”的字节流输出流，第6行以UTF－8字符编码方式将字节输出流转换为字符输出流，第7行将字符输出流转换为缓冲字符输出流以提高书写效率，第8行写入一行数据，第9行关闭缓冲字符输出流。

**案例**

模拟记事本，实现对文本文件的读入、修改、保存功能。

【分析】

（1）读入指定位置的文本文件：使用FileReader类实现文本文件的读取。

（2）多行文本框展示文本文件。

（3）保存文件：使用FileWriter类实现文本文件的保存。

**1. 窗体设计**

窗体设计如图 9－2－4 所示。

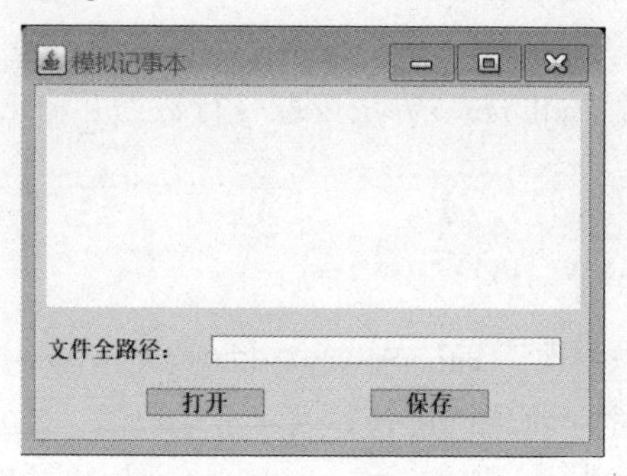


图 9－2－4　窗体设计

**2. 功能实现**

代码如下：

```
public class IOExercise {
    private JFrame frame;
    private JTextField textField;
    String filepath;
    ……(略)
    private void IOExercise () {
        ……(略)
        JButton btnNewButton = new JButton("打开");
        btnNewButton.addActionListener(new ActionListener() {
            public void actionPerformed(ActionEvent arg0) {
                filepath = textField.getText();
                File f = new File(filepath);
                FileReader fr = null ;
                BufferedReader bfr = null ;
                try {
                    fr = new FileReader(f);
                    bfr = new BufferedReader(fr);
                    String s = null ;
                    String display = null ;
                    while ((s = bfr.readLine())! = null ){
                        display = display + s + "\r\n";
                        textArea.setText(display);;
                    }
                } catch (Exception e) {
                        e.printStackTrace();
                }finally {
                    try {
                        bfr.close();
                    } catch (IOException e) {
                        e.printStackTrace();
                    }
```

```
                }
            }
        });
        ……(略)
        JButton btnNewButton_1 = new JButton("保存");
        btnNewButton_1.addActionListener(new ActionListener() {
            public void actionPerformed(ActionEvent arg0) {
                filepath = textField.getText();
                File f = new File(filepath);
                FileWriter fw = null;
                BufferedWriter bfw = null;
                try {
                    fw = new FileWriter(f);
                    bfw = new BufferedWriter(fw);
                    bfw.write(textArea.getText());
                } catch (Exception e) {
                    e.printStackTrace();
                }finally {
                    try {
                        bfw.close();
                    } catch (IOException e) {
                        e.printStackTrace();
                    }
                }
            }
        });
    ……(略)
}
```

## 任务工单

<table>
<tr><td>任务名称</td><td colspan="5">图书信息管理系统——从“books. txt”文件读取图书信息</td></tr>
<tr><td>班级</td><td></td><td>学号</td><td></td><td>姓名</td><td></td></tr>
<tr><td>任务要求</td><td colspan="5">读入图书信息<br>从books.txt文件读取图书信息<br>Book [bookID=001, bookName=Java程序设计, writer=王强, press=清华大学出版社, price=50.0, status=在库]<br>Book [bookID=002, bookName=Java案例教程, writer=王新, press=北京大学出版社, price=58.0, status=在库]<br>Book [bookID=003, bookName=Java程序设计从入门到精通, writer=周北平, press=清华大学出版社, price=46.0, status=在库]<br>Book [bookID=004, bookName=C语言程序设计, writer=何贝, press=高等教育出版社, price=38.0, status=在库]<br>Book [bookID=005, bookName=C语言案例教程, writer=周华金, press=高等教育出版社, price=40.0, status=已借]<br>Book [bookID=006, bookName=Python程序设计, writer=林琳, press=北京大学出版社, price=50.0, status=已借]<br><br>使用字符流，读取 D 盘中“books. txt”文件（d：\books. txt）中的图书信息并显示。</td></tr>
<tr><td>任务实现</td><td colspan="5"></td></tr>
<tr><td>异常及待解决问题记录</td><td colspan="5"></td></tr>
</table>

## 练习题

### 1. 选择题

(1) 下列关于在 Java 中删除目录的说法错误的是（　　）。

A. 在 Java 中删除目录是从 Java 虚拟机中直接删除（不放入回收站），文件无法恢复

B. File 类的 delete( )方法不允许将带有子文件的目录直接删除

C. File 类的 delete( )方法可以删除一个指定的文件

D. 在删除目录时，先删除这个目录，再删除该目录下的所有文件

(2) 下列选项中，哪个不是 InputStream 类的直接子类？(　　)

A. ByteArrayInputStream

B. FileInputStream

C. BufferedInputStream

D. PipedInputStream

(3) 下列选项中，哪个是 FileInputStream 类的父类？(　　)

A. File　　B. FileOutput　　C. OutputStream　　D. InputStream

(4) 为了从文本文件中逐行读取内容，应该使用（　　）类处理流对象。

A. BufferedWriter　　B. BufferedReader

C. BufferedOutputStream　　D. BufferedInputStream

(5) 为了实现自定义对象的串行化，该自定义对象必须实现（　　）接口。

A. Runnable　　B. Transient　　C. Volatile　　D. Serializable

(6) 在输入流的 read( )方法返回（　　）值表示读取结束。

A. null　　B. 1　　C. -1　　D. 0

(7) 下列关于 java. io. FileOutputStream 的说法错误的是（　　）。

A. java. io. FileOutputStream 用于进行文件的写入操作。用它提供的方法可以将指定文件写入本地主机硬盘

B. 通过 File 类的实例或者一个表示文件名称的字符串可以生成文件输出流。在流对象生成的同时文件被打开，但还不能进行文件读写

C. java. io. FileOutputStream 既可以读取字节文件，也可以读取字符文件

D. java. io. FileOutputStream 一次只能读取一个字节的内容

(8) 阅读下列代码。

```
import java.io. * ;
public class Example{
public static void main(String[] args) throws Exception {
OutputStream out = new FileOutputStream("itcast.txt ", true);
String str = "欢迎你!";
byte[] b = str.getBytes();
for (int i = 0; i < b.length; i ++) {
        out. _______(b[i]);
    }
    out.close();
```

```
    }
}
```

上述代码中的横线上应填写的方法名称是（　　）。

A. read()　　　　B. write()

C. close()　　　　D. available()

**2. 填空题**

（1）字符流有两个抽象的顶级父类，分别是 Reader 和__________。

（2）IO 流有很多种，按照操作数据的不同，可以分为__________和字符流。

（3）InputStream 类中专门用于读取文件中数据的子类是__________。

（4）JDK 提供的两个转换流类分别是__________和 OutputStreamWriter。

**3. 编程题**

（1）D 盘中的“article. txt”文件是一篇英文短文，编写一个程序，统计该文件中英文字母的个数，并将其写入另一个文本文件。

（2）定义学生类，包含学生的基本信息（包括学号、姓名和考试成绩），从键盘接收学生信息，并将学生信息保存到文件“studentscore. obj”中。

第 10 章

# 多线程编程

——勇往直前，砥砺前行

# 10.1 线程及其生命周期

## 学习目标

（1）理解线程的概念。

（2）掌握继承 Thread 类创建多线程的方法。

（3）掌握实现 Runnable 接口创建多线程的方法。

（4）掌握线程的生命周期及状态转换。

（5）理解线程的优先级。

（6）掌握线程的休眠、插队、让步。

## 相关知识

### 10.1.1 多线程概述

在计算机中，采用多线程技术同时执行多个任务。Java 内置了对多线程技术的支持，可以使程序同时执行多个执行片段。

#### 一、进程

在学习线程之前，需要首先了解什么是进程。在操作系统中，每个正在运行的程序都可称为一个进程。在 Windows 操作系统中，在“任务管理器”中可以看到当前正在运行的所有程序，也就是操作系统中的所有进程，如图 10－1－1 所示。

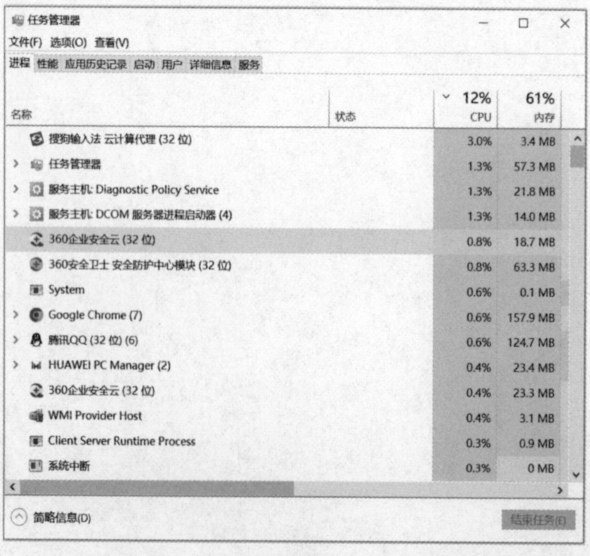

图 10－1－1 “任务管理器”中的进程管理

所有进程好像在同时运行，事实上对于单 CPU 来说，在某个时间点只运行一个进程。因此，计算机将 CPU 的使用时间分成小片分配给每个进程，如果某个进程获得时间片，则 CPU 运行该进程，当时间片使用结束时，该进程则停止运行，CPU 继续运行下一个获得时间片的进程。由于 CPU 的运行速度非常高，能在很短的时间内自如地进行进程切换，所以人感觉不到进程切换，误认为多个进程是同时运行的。

### 二、线程

每个运行的程序都是一个进程，但进程并不是最小的执行单元，一个进程还可以有多个执行单元同时运行，这些执行单元称为线程。操作系统中的每个进程中都至少存在一个线程。例如，当一个 Java 程序启动时，就会产生一个进程，该进程会默认创建一个线程，在这个线程上会运行 main( ) 方法中的代码。

**1. 单线程程序**

如图 10－1－2（a）所示，一个进程运行时只产生一个线程（类似单人完成一项工作），程序段按照调用顺序依次往下运行，不会出现两个程序段交替运行的情况，这样的程序称为单线程程序。

**2. 多线程程序**

如图 10－1－2（b）所示，一个进程运行时产生多个线程（类似多人合作完成一项工作），这些线程在运行时是相互独立的，它们可以同时运行不同的程序段，因此就出现多个程序段交替运行的情况，这样的程序称多线程程序。

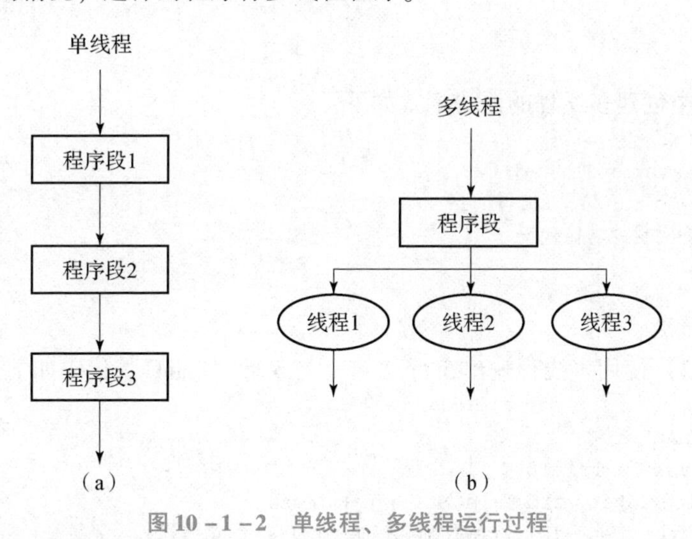


**图 10－1－2　单线程、多线程运行过程**

（a）单线程；（b）多线程

## 10.1.2　线程的创建

Java 提供了两种多线程实现方式，一种是继承 java. lang 包下的 Thread 类，覆写 Thread 类的 run( ) 方法，将线程要实现的功能代码写在 run( ) 方法中；另一种是实现 java. lang. Runnable 接口，同样将要实现的功能代码写在 run( ) 方法中。

【例 10-1-1】单线程程序输出 5 行“*”与 5 行“hello”。代码如下：

```
1  public class Example10_1_1 {
2      public static void main(String[] args) {
3          ThreadDemo th1 = new ThreadDemo ();//新建 Thread 类对象
4              th1.run();//调用 Thread 类的 run()方法
5              for (int i =1;i <=5;i ++)
6                  System.out.println("main()函数" + " ----第" + i + "个 hello");
7      }
8  }
9  class ThreadDemo{
10     public void run() {
11         for (int i =1;i <=5;i ++)
12             System.out.println("* * * * * * * * * * * *");
13     }
14 }
```

运行结果如图 10-1-3 所示。

```
***********
***********
***********
***********
***********
main()函数----第1个hello
main()函数----第2个hello
main()函数----第3个hello
main()函数----第4个hello
main()函数----第5个hello
```

图 10-1-3　运行结果

【说明】

程序先执行了 Thread 类的 run()方法输出 5 行“*”，然后执行 main()函数的循环语句输出 5 个“hello”，严格按照程序的调用顺序执行。

## 一、继承 Thread 类创建多线程

继承 Thread 类创建多线程的语法格式如下：

```
class  线程名 extends Thread{
    public void run(){
      //线程要实现的功能代码
    }
}
```

【例 10-1-2】使用多线程输出 5 行“*”与 5 行“hello”。代码如下：

```
1  package chapter10;
2  public class Example10_1_2 {
3      public static void main(String[] args) {
4          ThreadDemo th1 = new ThreadDemo();//创建线程
5          th1.start();//启动线程
6          for (int i =1;i <=5;i ++)
7              System.out.println("main()函数" + " ----第" + i + "个 hello");
8      }
9  }
10  class ThreadDemo extends Thread{
11      public void run() {
```

```
12          for (int i =1;i <=5;i ++)
13              System. out. println("* * * * * * * * * * * *");
14      }
15  }
```

运行结果如图 10 - 1 - 4 所示。

【说明】

程序第 10 行 ThreadDemo 继承 Thread 类成为线程类，第 4 行创建线程，第 5 行调用线程的 start( ) 方法启动线程，以线程方式执行 run( ) 方法。从运行结果看，“ * ” 行与 “hello” 行交替输出，这也说明在多线程中，main( ) 方法和 ThreadDemo 类的 run( ) 方法是同时运行的，互不影响，这正是单线程和多线程的区别。

```
main()函数----第1个hello
main()函数----第2个hello
main()函数----第3个hello
***********
***********
***********
***********
***********
main()函数----第4个hello
main()函数----第5个hello
```

图 10 - 1 - 4　运行结果

## 二、实现 Runnable 接口创建多线程

通过继承 Thread 类创建多线程具有一定的局限性，如果需要继承其他类（非 Thread 类)，还要使当前类实现多线程，那么可以通过 Runnable 接口来实现。例如，一个扩展 JFrame 类的 GUI 程序不可能再继承 Thread 类，因为 Java 不支持多继承，这时就需要实现 Runnable 接口使其具有使用线程的功能。Thread 类提供了另外一个构造方法 Thread (Runnable target)，其中，Runnable 是一个接口，该接口包含一个抽象的 run( ) 方法。当通过 Thread( Runnable target) 构造方法创建线程对象时，只需要为该构造方法传递一个实现了 Runnable 接口的实例对象，这样创建的线程将调用实现了 Runnable 接口的类中的 run( ) 方法作为运行代码，而不需要调用 Thread 类中的 run( ) 方法。实现 Runnable 接口创建多线程的语法格式如下：

```
class  实现类 implements  Runnable{
  public void run(){
       //线程要实现的功能代码
  }
}
Thread  线程名 =new  Thread(实现类对象);
```

【例 10 - 1 - 3】使用多线程输出 5 行 “hello”。代码如下：

```
1  package chapter10;
2  public class Example10_1_3 implements Runnable{
3      @ Override
4      public void run() {
5          for (int i =1;i <=5;i ++)
6              System. out. println("线程
                       " +Thread.currentThread().getName() +" ---hello");
```

```
7       }
8       public static void main(String[] args) {
9           Example10_3 ex10_3 = new Example10_3();
10          Thread th1 = new Thread(ex10_3);
11          Thread th2 = new Thread(ex10_3);
12          th1.start();
13          th2.start();
14      }
15  }
```

运行结果如图 10 - 1 - 5 所示。

【说明】

程序第 2 行的 Example10_1_3 类实现了 Runnable 接口，并覆写了 run( )方法；第 9 行定义了 Example10_1_3 类的对象，该对象作为线程的参数；第 10、11 行创建线程；第 12、13 行启动线程。

```
线程Thread-0---hello
线程Thread-0---hello
线程Thread-0---hello
线程Thread-1---hello
线程Thread-1---hello
线程Thread-1---hello
线程Thread-1---hello
线程Thread-1---hello
线程Thread 0---hello
线程Thread-0---hello
```

图 10 - 1 - 5　运行结果

### 10.1.3　线程的生命周期

线程具有生命周期，其中包含 5 种状态，分别为新建状态、就绪状态、运行状态、阻塞状态和死亡状态，如图 10 - 1 - 6 所示。

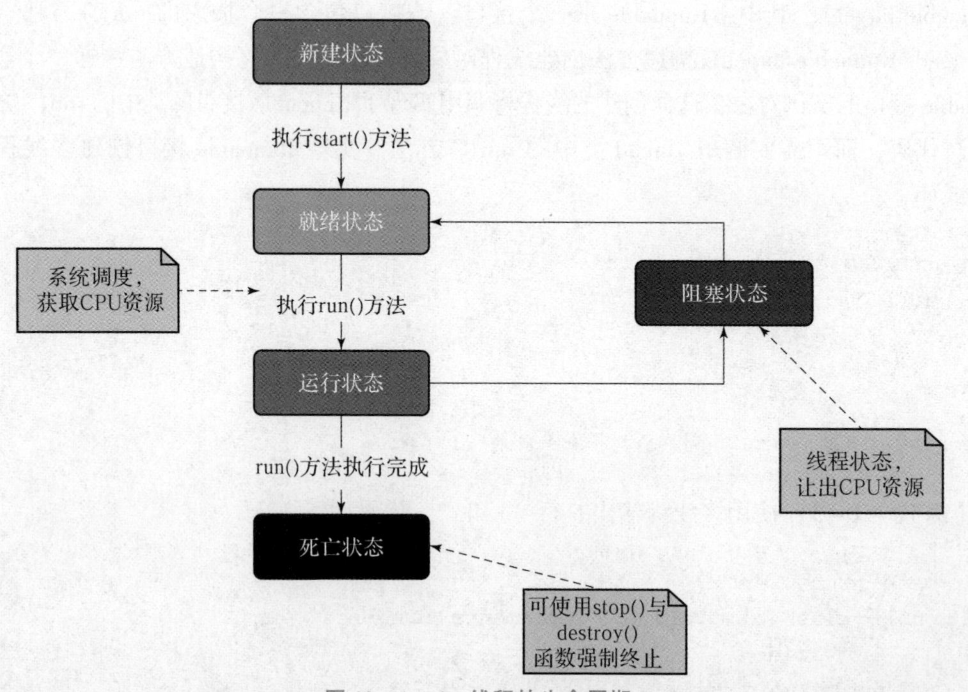


图 10 - 1 - 6　线程的生命周期

（1）新建状态。创建一个线程对象后，该线程对象就处于新建状态，此时不能运行。

（2）就绪状态。线程对象调用 start( )方法后，线程进入就绪状态，位于线程队列中。

（3）运行状态。如果处于就绪状态的线程获得了 CPU 的使用权，则执行 run( )方法中的线程执行体，该线程处于运行状态。一个线程启动后，它可能不会一直处于运行状态，当处于运行状态的线程使用完系统分配的时间后，系统就会剥夺该线程占用的 CPU 资源，让其他线程获得运行的机会。

（4）阻塞状态。如果正在运行的线程被人为挂起或执行耗时的输入/输出操作时，会让出 CPU 的使用权并暂时中止自己的运行，进入阻塞状态。线程进入阻塞状态后，就不能进入线程队列。只有当引起阻塞的原因被消除后，线程才可以转入就绪状态。

（5）死亡状态。当线程调用 stop( )方法或 run( )方法正常执行完毕，或者线程抛出一个未捕获的异常（Exception）、错误（Error），线程就进入死亡状态。一旦进入死亡状态，线程将不再拥有运行的资格，也不能再转换到其他状态。

### 10.1.4　线程的调度

多线程看起来像同时运行，但事实上在同一时间点只有一个线程运行，只是线程之间切换较快，使人产生线程同时运行的错觉。线程若想运行，必须得到 CPU 的使用权，Java 虚拟机会按照特定的机制为程序中的每个线程分配 CPU 的使用权，这种机制称为线程的调度。线程的调度有两种方式：分时调度及抢占调度。分时调度是指所有线程轮流获得相同时间片的 CPU 的使用权；抢占调度是指运行池中优先级高的线程优先占用 CPU，对于优先级相同的线程，则随机选择一个线程使其占用 CPU。Java 虚拟机默认采用抢占调度方式，当有特定需求时可以改变这种方式，由程序自己来控制线程的调度。

#### 一、线程的优先级

Java 虚拟机默认采用抢占调度方式，并赋予每个线程相应的优先级。优先级表明线程的重要性，优先级越高的线程获得 CPU 的使用权的机会越大，但这并不意味着优先级低的线程得不到运行的机会，而只是它运行的机会比较小，例如垃圾回收线程的优先级就较低。

线程的优先级采用 1 ~ 10 的整数表示，数字越大优先级越高；也可以使用 Thread 类的 3 个静态常量表示线程的优先级。

（1）static int MAX_PRIORITY：表示线程的最高优先级，值为 10。

（2）static int MIN_PRIORITY：表示线程的最低优先级，值为 1。

（3）static int NORM_PRIORITY：表示线程的普通优先级，值为 5。

在程序运行期间，处于就绪状态的每个线程都有自己的优先级，但线程的优先级不是固定不变的，可以通过 Thread 类的 setPriority(int newPriority) 方法对其进行设置，该方法中的参数 newPriority 为 1 ~ 10 的整数或者 Thread 类的 3 个静态常量。

**【例 10 - 1 - 4】** 代码如下：

```
1  public class Example10_1_4 {
2      public static void main(String[ ] args) {
3          ThreadDemo5 th0 = new ThreadDemo5();
```

```
4        ThreadDemo5 th1 = new ThreadDemo5();
5        Th0.setPriority(Thread. MAX_PRIORITY );//优先级为 10
6        Th1.setPriority(Thread. MIN_PRIORITY );//优先级为 1
7        Th0.start();
8        Th1.start();
9    }
10 }
11 class ThreadDemo5 extends Thread{
12     public void run() {
13         for (int i =1;i <=5;)
14             System. out. println(Thread.currentThread().getName() + " -- 第
                                                    " +(i ++) + "个 hello");
15         }
16 }
```

运行结果如图 10－1－7 所示。

## 二、操作线程的方法

```
Thread-0--第1个hello
Thread-1--第1个hello
Thread-0--第2个hello
Thread-0--第3个hello
Thread-1--第2个hello
Thread-0--第4个hello
Thread-0--第5个hello
Thread-1--第3个hello
Thread-1--第4个hello
Thread-1--第5个hello
```

图 10－1－7　运行结果

操作线程有很多方法，这些方法可以使线程从某种状态过渡到另一种状态。线程进入就绪状态的方法有 3 种：调用 sleep( )方法、调用 wait( )方法、等待输入/输出完成。线程进入运行状态的方法有 4 种：调用 notify( )方法、调用 notifyAll( )方法、等待线程的休眠时间结束、等待输入/输出完成。

### 1. 线程的休眠

为了使正在运行的线程暂停，将 CPU 让给其他线程，可以人为地调用能控制线程行为的方法 sleep( )，sleep( )方法使线程进入休眠状态，传入参数用于指定该线程休眠的时间，单位为毫秒。当休眠时间结束后，线程返回就绪状态，而不是立即开始运行。

### 2. 线程的插队

在多线程程序中，已存在线程 A，现在需要插入线程 B，并要求线程 B 先运行，然后线程 A 运行，这种插队行为可以使用 Thread 类中的 join( )方法实现。

### 3. 线程的让步

Thread 类提供了一种让步方法 yield( )，使用 yield( )方法可以让当前正在运行的线程暂停，将线程转换至就绪状态，让系统的调度器重新调度一次。当某个线程调用 yield( )方法之后，只有与当前线程优先级相同或者更高的线程才能获得运行的机会。

**案例**

火车站有 3 个售票窗口同时开放售票，假设此时还有 100 张火车票，3 个售票窗口如何实现同时售票？

【分析】

使用多线程实现 3 个售票窗口同时售票，实现方式如下。

（1）继承 Thread 类。

（2）实现 Runnable 接口。

1. 使用继承 Thread 类的方式实现同时售票

代码如下：

```
1  public class Ticket01 extends Thread{
2      int tick = 100;
3      public void run(){
4          while (true){
5              if (tick > 0) {
6              System. out. println(Thread.currentThread().getName() +"sale—" +tick);
7                          tick --;
8                  }
9              }
10         }
11 }
12 public class Test{
13     public static void main(String[] args) {
14         Ticket01 th1 = new Ticket01();
15         th1.start();
16         Ticket01 th2 = new Ticket01();
17         th2.start();
18         Ticket01 th3 = new Ticket01();
19         th3.start();
20     }
21 }
```

运行结果如图 10－1－8 所示。

```
Thread-0sale--100
Thread-0sale--99
Thread-0sale--98
Thread-0sale--97
Thread-1sale--100
Thread-1sale--99
Thread-2sale--100
Thread-2sale--99
```

图 10－1－8　运行结果

【说明】

从运行结果可以看出，同一张火车票被卖出 3 次，出现该情况的原因是 100 张火车票被定义在 Ticket01 类中，在创建 3 个线程时，Ticket01 类被实例化 3 次，因此就有 300 张火车票被卖出。

2. 使用实现 Runnable 接口的方式实现同时售票

代码如下：

```
1  public class Ticket02 implements Runnable{
2      int tick = 100;
3      @ Override
4      public void run() {
```

```
5          while(true){
6             if(tick>0){
7     System.out.println(Thread.currentThread().getName()+"sale--"+tick);
8                 tick--;
9             }
10          }
11       }
12    }
13    public class Test{
14       public static void main(String[] args){
15          Ticket02 tick = new Ticket02();
16          Thread th1 = new Thread(tick);
17          Thread th2 = new Thread(tick);
18          Thread th3 = new Thread(tick);
19          th1.start();
20          th2.start();
21          th3.start();
22       }
23    }
```

运行结果如图 10-1-9 所示。

```
Thread-1sale--100
Thread-1sale--99
Thread-1sale--98
Thread-1sale--97
Thread-1sale--96
Thread-1sale--95
Thread-1sale--94
Thread-2sale--100
Thread-2sale--92
```

图 10-1-9　运行结果

【说明】

从运行结果可以看出，这种实现方式解决了同时卖出 300 张火车票的问题，但仍然存在一票多卖的情况。10.2 节将解决这个问题。

# 10.2 线程安全与线程同步

## 学习目标

（1）理解线程安全。
（2）掌握同步块的使用方法。
（3）掌握同步方法。
（4）理解死锁。

## 相关知识

单线程按先后顺序运行，多线程的并发运行可以提高程序的效率，但会发生两个线程抢占资源引发的安全问题。因此，Java 提供了线程同步的机制来防止多线程编程中资源访问冲突，即限制某个资源在同一时刻只能被一个线程访问。

### 10.2.1 线程安全

多线程程序在实际开发中应用广泛，例如银行排队系统、售票系统等。多线程程序的安全问题在售票系统中体现为"一票多卖""超额卖票"等情况。程序只要判断当前火车票数大于 0，就将火车票出售给乘客，但当两个线程同时访问这段程序时，假设火车票只剩 1 张，在第一个线程将火车票售出的同时，第二个线程也已经完成余票判断，并得出火车票数大于 0 的结论，于是第二个线程也执行售票操作，这样就会产生负数票。因此，在编写多线程程序时，应该考虑到线程安全问题。实质上线程安全问题来源于两个线程同时存取单一对象的数据。

修改 10.1 节中的案例代码，在售票的代码中使用 sleep( )方法，每次售票后线程休眠 5 毫秒。代码如下：

```
1  public class Ticket02 implements Runnable{
2      int tick =100;
3      @ Override
4      public void run() {
5          while (tick >0){
6  System. out. println(Thread.currentThread().getName() +"sale -- " +tick);
7              tick --;
8          }
9      }
10 }
11  public class Test{
12      public static void main(String[] args) {
13          Ticket02 tick =new Ticket02();
14          Thread th1 =new Thread(tick);
15          Thread th2 =new Thread(tick);
16          Thread th3 =new Thread(tick);
17          th1.start();
18          th2.start();
```

```
19          th3.start();
20      }
21  }
```

运行结果如图 10 - 2 - 1 所示。

```
Thread-1sale--100
Thread-1sale--99
Thread-1sale--98
Thread-1sale--97
Thread-1sale--96
Thread-1sale--95
Thread-1sale--94
Thread-2sale--100
Thread-2sale--92
```

图 10 - 2 - 1　运行结果

### 10.2.2　线程同步

如何解决资源共享引发的安全问题？Java 采用共享资源上锁的方法，保证在给定时间只允许一个线程访问共享资源。这就好比一个人在使用洗手间时会将门锁上，使用完毕再将锁打开，然后其他人才可以使用洗手间。

#### 一、同步块

Java 提供了同步机制，可以有效地防止资源冲突。同步机制使用 synchronized 关键字，其语法格式如下：

```
synchronized(obj){
   //同步块中的语句
}
```

修改 10.1 节中的案例代码，使用同步块锁定一次售票代码，直到该代码块中的所有代码都被执行后再释放，允许出售下一张火车票。代码如下：

```
1   public class Ticket011 implements Runnable{
2       int tick =100;
3       Object obj = new Object();
4       @ Override
5       public void run() {
6           while (true){
7               synchronized (obj) {
8                   if (tick >0){
9       System. out. println(Thread.currentThread().getName() + "sale -------" +
        tick --);
10                      try {
11                          Thread.sleep(100);
12                      } catch (InterruptedException e) {
13                          //TODO Auto - generated catch block
14                          e.printStackTrace();
```

```
15                    }
16                }
17            }
18        }
19    }
20}
```

运行结果如图 10－2－2 所示。

```
Thread-2sale-------5
Thread-0sale-------4
Thread-0sale-------3
Thread-2sale-------2
Thread-1sale-------1
```

图 10－2－2　运行结果

## 二、同步方法

同步方法就是在方法前面加 synchronized 关键字的方法，其语法格式如下：

```
synchronized  void  f(){}
```

当某个对象调用了同步方法时，该对象上的其他同步方法必须等待该同步方法执行完毕后才能被执行。必须将每个能访问共享资源的方法用 synchronized 关键字修饰，否则会出错。例如：

```
1  public class TicketThread implements Runnable{
2      int tick =100;
3      public synchronized  void  saleTicket() {
4          if(tick >0){
5  System.out.println(Thread.currentThread().getName() +"sale------" +tick --);
6          }
7          try {
8              Thread.sleep(100);
9          } catch (InterruptedException e) {
10             //TODO Auto-generated catch block
11             e.printStackTrace();
12         }
13     }
14     public void run() {
15         while(true)
16             saleTicket();
17     }
18 }
```

运行结果如图 10－2－3 所示。

```
Thread-2sale-------6
Thread-2sale-------5
Thread-2sale-------4
Thread-2sale-------3
Thread-1sale-------2
Thread-1sale-------1
```

图 10－2－3　运行结果

将共享资源的操作放置在同步方法中，运行结果与使用同步块的结果一致。

**案例**

设计一个秒表，如图 10－2－4 所示。

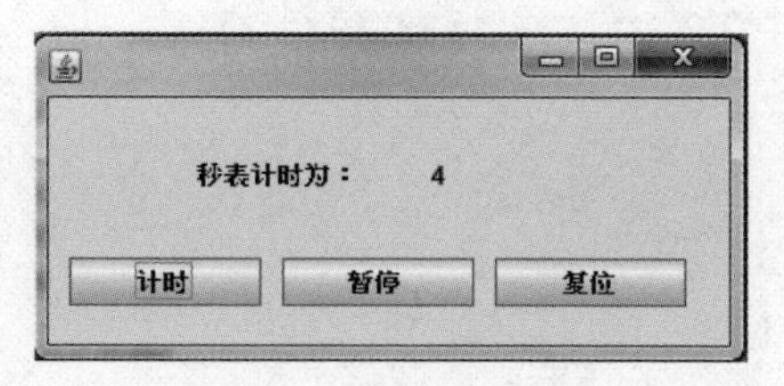


图 10-2-4　秒表

【分析】

使用多线程实现秒表，其功能如下。

（1）单击“计时”按钮，则秒表开始计时，每秒增加1。

（2）单击“暂停”按钮，则秒表计时暂停，直到单击“计时”按钮再继续计时。

（3）单击“复位”按钮，则秒表计时清零。

**1. 秒表窗体**

代码如下：

```
public class StopWatch extends JFrame {
    private JPanel contentPane;
    JLabel lblNewLabel_1;
    int n;
    StopWatchThread swt;
    Thread thread;
    public int getN() {
        return n;
    }
    public void setN(int n) {
        lblNewLabel_1.setText("" +n);
    }
    ……(略)
    public StopWatch() {
        ……(略)
        swt =new StopWatchThread();
        JButton button  = new JButton("计时");
        button.addActionListener(new ActionListener() {
              public void actionPerformed(ActionEvent e) {
              thread =new Thread(swt);
              swt.setFlag(true );
              swt.setSw(StopWatch. this );
              thread.start();
        }
    });
    button.setBounds(10, 76, 93, 23);
    contentPane.add(button);
    JButton button_1  = new JButton("暂停");
    button_1.addActionListener(new ActionListener() {
```

```
        public void actionPerformed(ActionEvent e) {
            swt.setFlag(false);
        }
    }
    button_1.setBounds(113, 76, 93, 23);
    contentPane.add(button_1);
    JButton button_2 = new JButton("复位");
    button_2.addActionListener(new ActionListener() {
        public void actionPerformed(ActionEvent e) {
            swt.setN(0);
            lblNewLabel_1.setText("" +0);
        }
    });
    ……(略)
}
```

### 2. 使用线程实现秒表的“计时”“暂停”“复位”功能

代码如下：

```
public class StopWatchThread implements Runnable{
    boolean flag = false;
    StopWatch sw;
    Object obj = new Object();
    int n = 0;
    //省略 get()、set()方法
    public void run() {
        while(true){
            synchronized(obj){
                if(flag){
                    String s = sw.lblNewLabel_1.getText();
                     n = Integer.parseInt(s);
                     n ++;
                     sw.setN(n);
                     try {
                         Thread.sleep(1000);
                     } catch (InterruptedException e) {
                         e.printStackTrace();
                     }
                }
            }
        }
    }
}
```

## 任务工单

<table>
<tr><td>任务名称</td><td colspan="6">图书信息管理系统——赠书游戏</td></tr>
<tr><td>班级</td><td></td><td>学号</td><td></td><td>姓名</td><td></td></tr>
<tr><td>任务要求</td><td colspan="6"> 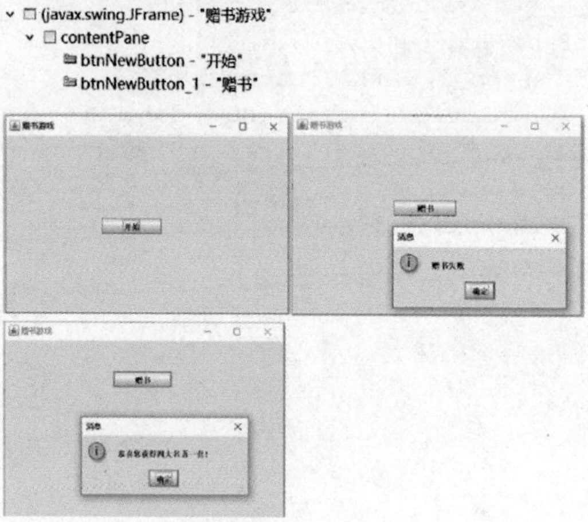(javax.swing.JFrame) - "赠书游戏"<br>contentPane<br>btnNewButton - "开始"<br>btnNewButton_1 - "赠书"<br><br>（1）单击“开始”按钮，赠书游戏开始，此时“开始”按钮隐藏，“赠书”按钮随机改变位置。<br>（2）单击“赠书”按钮，给出提示“恭喜您获得四大名著一套!”，赠书游戏结束。<br>（3）当“赠书”按钮随机改变位置 20 次后，用户还没有捕捉到“赠书”按钮，则赠书失败，赠书游戏结束。</td></tr>
</table>

续表

<table>
<tr><td>任务名称</td><td colspan="5">图书信息管理系统——赠书游戏</td></tr>
<tr><td>班级</td><td></td><td>学号</td><td></td><td>姓名</td><td></td></tr>
<tr><td>任务实现</td><td colspan="5">窗体代码如下：<pre>
public class Game extends JFrame {



public Game() {
  btnNewButton = new JButton("开始");
  btnNewButton.addActionListener(new ActionListener() {
  public void actionPerformed(ActionEvent e) {



}
}
  ……(略)
  btnNewButton = new JButton("赠书");
  btnNewButton.addActionListener(new ActionListener() {
  public void actionPerformed(ActionEvent e) {



}
}
  }
}
</pre>线程代码如下：</td></tr>
<tr><td>异常及待解决问题记录</td><td colspan="5"></td></tr>
</table>

## 练习题

### 1. 选择题

(1) 启动线程的方法是（　　）。

A. join( )　　B. run( )　　C. start( )　　D. sleep( )

(2) 创建线程时必须实现（　　）接口。

A. Runnable　　B. Thread　　C. Run　　D. Start

(3) 下列关于同步块的特征的说法错误的是（　　）。

A. 可以解决线程安全问题

B. 会降低程序的性能

C. 使用 synchronized 关键字修饰

D. 多线程同步的锁只能是 object 对象

(4) 下列有关多线程中静态同步方法的说法正确的是（　　）。

A. 对于静态同步方法而言，该方法的同步监视器不是 this，而是该类本身

B. 在使用同步块时，静态同步方法既可以使用 class 对象同步，也可以使用 this 同步

C. 一个类中的多个静态同步方法可以同时被多个线程执行

D. 不同类中的静态同步方法被多线程访问时，线程需要等待

(5) 关于线程的死锁，下面的说法正确的是（　　）。

A. 若程序中存在线程的死锁问题，则编译时不能通过

B. 线程的死锁是一种逻辑运行错误，编译器无法检测

C. 实现多线程时，线程的死锁不可避免

D. 为了避免线程的死锁，应解除对资源以互斥的方式进行访问

(6) 当一个处于阻塞状态的线程解除阻塞状态后，它将回到（　　）。

A. 运行状态　　B. 结束状态

C. 新建状态　　D. 可运行状态

(7) 调用 Thread 类的 sleep( )方法后，当前线程（　　）。

A. 由运行状态进入阻塞状态　　B. 由运行状态进入等待状态

C. 由阻塞状态进入等待状态　　D. 由阻塞状态进入运行状态

### 2. 填空题

(1) 线程的状态主要有新建状态、________、________、________ 和结束状态 5 种。

(2) 当多个线程使用同一个共享资源时，可以将处理共享资源的代码放置在一个代码块中，使用________ 关键字修饰，称为同步块。

(3) 多线程中出现多个线程循环等待他方占用的资源而无限期地僵持下去的局面称为__________。

(4) Java 中有两种创建线程的方式，一种是继承 Thread 类，另一种是实现________接口。

(5) 执行________ 方法，可以让线程在规定的时间内休眠。

（6）在 Java 中，线程的让步可以通过________ 方法来实现。

（7）在 Java 中，线程间的通信通过____________和____________方法进行。

**3. 编程题**

（1）编写一个程序，启动 3 个线程，3 个线程的 ID 分别是 A，B，C。每个线程将自己的 ID 在屏幕上打印 5 遍，打印顺序是“ABCABC…”。

（2）编写一个程序，启动 2 个线程，一个线程打印 1～52，另一个线程打印 A～Z，打印顺序是“12A34B…51Y52Z”。

# 第11章 JDBC编程

——学而不厌，诲人不倦

# 11.1 JDBC 概述

## 学习目标

（1）了解 JDBC 的概念。

（2）了解 JDBC 驱动程序。

（3）了解 JDBC 常用 API。

## 相关知识

### 11.1.1 JDBC 的概念

JDBC（Java DataBase Connectivity）是 Java 操作数据库的技术规范。JDBC 定义了一组访问和操作关系数据库的类与接口，要实现对数据库的操作就必须实现这些接口。JDBC 接口包含两个部分：一个是顶层面向应用的 API，即 Java API，负责在 Java 程序和 JDBC 驱动程序管理器之间进行通信，实现数据库的连接、执行 SQL 语句、获得执行结果等；另一个是底层面向数据库的 API，即 Java Driver API，负责 JDBC 驱动程序管理器与实际连接的数据库的厂商驱动程序和第三方驱动程序之间进行通信，返回查询信息或者执行规定的操作。图 11-1-1 所示为 JDBC 访问数据库功能示意。

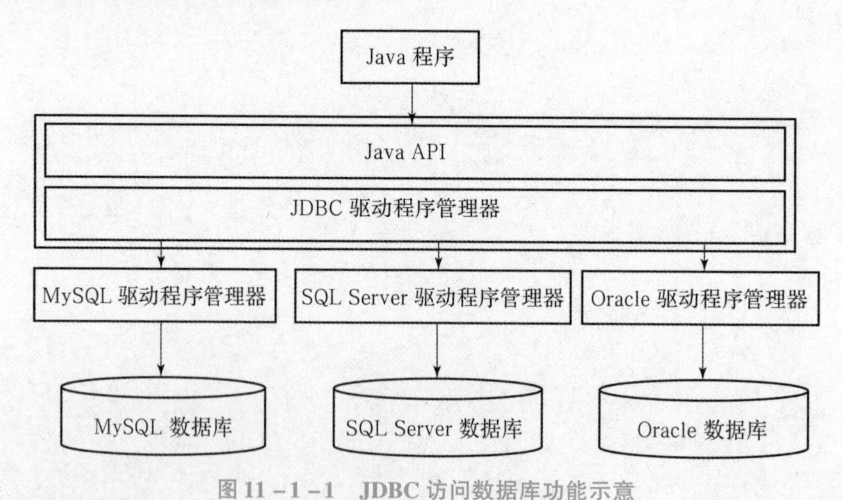


图 11-1-1　JDBC 访问数据库功能示意

### 11.1.2 JDBC 驱动程序

不同的数据库厂商为了能让 Java 使用自己的数据库都提供了自己的 JDBC 驱动程序，也就是说，Java 只要根据所使用数据库类型加载相应的 JDBC 驱动程序即可。JDBC 驱动程序有以下 4 种。

（1）JDBC－ODBC 桥接驱动程序。

（2）本地 API 驱动程序。

（3）网络协议驱动程序。

（4）本地协议纯 Java 驱动程序。

不管使用以上哪种 JDBC 驱动程序，数据库的连接方式都是相同的，都必须加载选定类型数据库的驱动程序，利用该驱动程序创建与数据库的连接，然后通过相关类和接口访问、操作数据库。常用的数据库驱动程序如表 11－1－1 所示。

表 11－1－1　常用的数据库驱动程序

| 数据库名 | 驱动程序名 | jar 包 |
| --- | --- | --- |
| ODBC 数据源 | sun. jdbc. odbc. JdbcOdbcDriver | 不需要 jar 包，直接配置数据源 |
| SQL Server | com. microsoft. sqlserver. jdbc. SQLServerDrvier | sqljdbc4. jar |
| MySQL | com. mysql. jdbc. Driver | mysql－connector－java－5. 1. 18－bin. jar |
| Oracle | oracle. jdbc. driver. OracleDriver | nls_charset12. jar、classes. jar |

## 11. 1. 3　JDBC 常用 API

JDBC API 位于 java. sql 包中，主要定义了一系列访问数据库的接口和类。

### 1. Driver 接口

Driver 接口是所有 JDBC 驱动程序必须实现的接口，该接口专门提供给数据库厂商使用。在程序中不直接访问实现了 Driver 接口的类，而是由驱动程序管理器类（java. sql. DriverManager）调用 Driver 接口实现类。

### 2. DriverManager 类

DriverManager 类负责加载各种不同的驱动程序（Driver），DriverManager 类的常用方法如表 11－1－2 所示。

表 11－1－2　DriverManager 类的常用方法

| 方法 | 功能描述 |
| --- | --- |
| static void registerDriver(Driver driver) | 向驱动程序管理器注册给定的 JDBC 驱动程序 |
| static Connection getConnection(String url，String user，String pwd) | 建立和数据库的连接，返回表示连接的 Connection 对象（url：指定要访问的数据库地址；user：指定连接数据库所使用的用户名；pwd：指定连接数据库所使用的密码） |

### 3. Connection 接口

Connection 接口表示与数据库的物理连接，负责与数据库的通信。SQL 语句的执行、事务的处理都是在某个特定的 Connection 接口中进行的。通过 Connection 对象可获取用于执行

SQL 语句的 Statement 接口对象、PreparedStatement 接口对象和 CallableStatement 接口对象等。Connection 接口的常用方法如表 11-1-3 所示。

表 11-1-3　Connection 接口的常用方法

| 方法 | 功能描述 |
| --- | --- |
| DatabaseMetaData getMetaData() | 返回表示数据库数据的 DatabaseMetaData 接口对象 |
| Statement createStatement() | 创建负责向数据库发送 SQL 请求的 Statement 接口对象 |
| PreparedStatement prepareStatement(String sql) | 创建负责向数据库发送 SQL 请求的 PreparedStatement 接口对象 |
| CallableStatement prepareCall(String sql) | 创建 CallableStatement 对象以调用数据库存储过程 |

### 4. Statement 接口

Statement 接口用于执行静态 SQL 语句，并带回一个结果对象。Statement 接口对象可以通过 Connection 接口实例的 createStatement()方法获得，该对象把静态 SQL 语句发送到数据库中编译和执行，并带回数据库的处理结果。Statement 接口的常用方法如表 11-1-4 所示。

表 11-1-4　Statement 接口的常用方法

| 方法 | 功能描述 |
| --- | --- |
| boolean execute(String sql) | 执行 SQL 语句（包括增、删、改、查等），返回值为 boolean 型，若返回值为 true，则表示 SQL 语句执行成功，可通过 Statement 接口的 getResultSet()方法获得查询结果 |
| int executeUpdate(String sql) | 执行 insert、update、delete 三种 SQL 语句，返回值为 int 型，表示数据库中受该 SQL 语句影响的记录数 |
| ResultSet executeQuery(String sql) | 执行 SQL 语句中的 select 语句，返回查询结果集 ResultSet 接口对象 |

### 5. PreparedStatement 接口

Statement 接口主要执行静态 SQL 语句，通过变量及 SQL 语句的拼接完成动态 SQL 语句的执行，但拼接的数量多时，SQL 语句变得冗长、可读性差，存在安全方面的问题。因此，JDBC API 提供了扩展的 PreparedStatement 接口。PreparedStatement 接口是 Statement 接口的子接口，用于执行包含动态参数的 SQL 语句。该接口中的 SQL 语句使用占位符“?”代替参数，SQL 语句被执行前通过 setXxx()方法为 SQL 语句的参数赋值即可达到传参的目的。PreparedStatement 接口的常用方法如表 11-1-5 所示。

表 11－1－5　PreparedStatement 接口的常用方法

| 方法 | 功能描述 |
| --- | --- |
| int executeUpdate() | PreparedStatement 接口执行 DML 语句或者无返回值的 SQL 语句，返回值为 int 型，表示数据库中受该 SQL 语句影响的记录数 |
| ResultSet executeQuery() | PreparedStatement 接口执行 SQL 查询语句，返回查询结果集 ResultSet 接口对象 |
| void setlnt(int parameterindex, int x) | 将指定参数设置为给定的 int 型的值 |
| void setFloat(int parameterindex, float x) | 将指定参数设置为给定的 float 型的值 |
| void setString(int parameterindex, String x) | 将指定参数设置为给定的 String 型的值 |
| void setDate(int parameterindex, Date x) | 将指定参数设置为给定的 Date 型的值 |

### 6. ResultSet 接口

ResultSet 接口实例用于保存 JDBC 执行查询时返回的结果集，它由多行、多列数据构成，等同于一张二维表。在 ResultSet 接口内部有一个指向表格数据行的游标（或指针），ResultSet 接口对象初始化时，游标指在表格的第一行之前，调用 next()方法可将游标移动到下一行。如果下一行没有数据，则 next()方法返回 false。在应用程序中经常使用 next()方法作为 while 循环的条件来迭代 ResultSet 接口。ResultSet 接口的常用方法如表 11－1－6 所示。

表 11－1－6　ResultSet 接口的常用方法

| 方法 | 功能描述 |
| --- | --- |
| XXX getXXX(int columnlndex) | 获取指定索引字段的 XXX 型的值，XXX 表示数据类型，例如 String、int、double 等 |
| XXX getXXX(String columnName) | 获取指定字段名的 XXX 型的值，XXX 表示数据类型，例如 String、int、double 等 |
| boolean next() | 将游标从当前位置向下移一行 |
| boolean absolute(int row) | 将游标移动到此 ResultSet 接口对象的指定行 |
| void afterLast() | 将游标移动到此 ResultSet 接口对象的末尾，即最后一行之后 |
| void beforeFirst() | 将游标移动到此 ResultSet 接口对象的开头，即第一行之前 |
| boolean previous() | 将游标移动到此 ResultSet 接口对象的上一行 |
| boolean last() | 将游标移动到此 ResultSet 接口对象的最后一行 |

# 11.2 JDBC 应用

## 学习目标

（1）掌握 JDBC 连接数据库的方法。
（2）掌握 JDBC 操作访问数据库的方法。

## 相关知识

Java 使用 JDBC 进行数据库程序设计主要包括以下 3 个步骤。
（1）连接数据库。
（2）执行查询语句并处理查询结果。
（3）关闭与数据库的连接。

### 11.2.1 连接数据库

Java 程序连接数据库主要有两个步骤：设置驱动程序、建立与数据库的连接。本章主要使用 MySQL 数据库，学习前请先安装 MySQL 8.0。

#### 一、设置驱动程序

**1. 下载 MySQL JDBC Driver 驱动 jar 包**

为了连接 MySQL 数据库，首先要下载 Java 连接 MySQL 数据库的驱动 jar 包（mysql - connector - java - 5.1.18 - bin. jar），如图 11 - 2 - 1 所示。不同的 MySQL 版本需要下载不同的驱动 jar 包，请根据自己的 MySQL 版本选择相应的驱动 jar 包进行下载（下载地址：https：// dev. mysql. com/ downloads/ connector/ j/）。

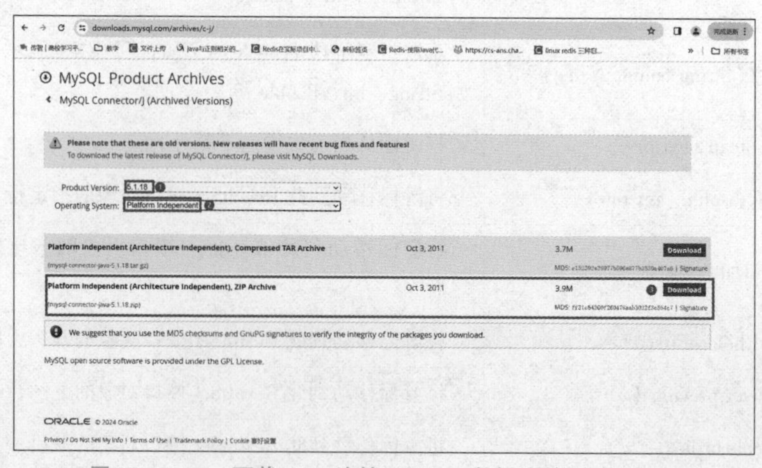


图 11 - 2 - 1 下载 Java 连接 MySQL 数据库的驱动 jar 包

### 2. 添加驱动 jar 包

将驱动 jar 包添加到工程的路径下，如图 11-2-2 所示。

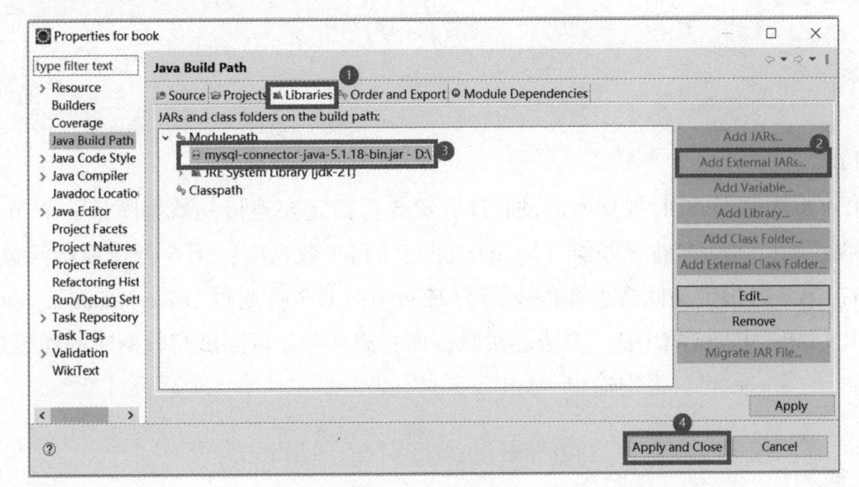


图 11-2-2　添加驱动 jar 包

### 3. 加载 MySQL 数据库的驱动程序

代码如下：

```
Class.forName("com.mysql.jdbc.Driver");
```

## 二、建立与数据库的连接

通过调用 DriverManager.getConnection(URL，数据库用户，密码）方法建立与数据库的连接，参数 URL 用于指定要访问的数据库的类型、地址、名称，JDBC 驱动程序管理器尝试使用“数据库用户”及“密码”连接 URL 字符串指定的数据库，若连接成功，则返回一个 Connection 接口对象。

**【例 11-2-1】** 代码如下：

```
1  public class P11_2_1 {
2      public static void main(String[] args) {
3          try {
4          Class.forName("com.mysql.jdbc.Driver");
5          String URL = "jdbc:mysql://localhost:3306/iot? useUnicode=true
                                               &characterEncoding=UTF-8";
6          String USERNAME = "root";
7          String PASSWORD = "123456";
8          Connection conn=DriverManager.getConnection(URL,USERNAME,
                                                          PASSWORD);
9          if(conn!=null)
10              System.out.println("success");
11         else
12              System.out.println("fail");
```

```
13            } catch (Exception e) {
14                e.printStackTrace();
15            }
16        }
17    }
```

【说明】

程序第 4 行加载 MySQL 数据库驱动程序；第 5 行指定要连接的数据库是一个 MySQL 数据库，通过 3306 端口访问本地服务器（localhost）上的 iot 数据库；第 6 行指定访问数据库的用户名为 root，第 7 行指定访问数据库的密码为 123456；第 8 行通过 DriverManager. getConnection (url, USERNAME, PASSWORD) 方法连接数据库；第 9～12 行测试与数据库的连接是否成功，若成功则输出“success”，否则输出“fail”。

### 11.2.2 执行查询语句并处理查询结果

#### 一、数据库准备

（1）新建数据库（javadb）。

（2）新建用户表（user）。

用户表包含 5 个字段，分别是 id（用户编号：int、自动增长、主键）、name（用户姓名：varchar(20)）、sex（性别：char(1)）、birthday（出生日期：date）、tel（联系电话：char(11)）。

（3）添加用户信息。

代码如下：

```
insert into user values (null,'杨洋','男','1990-10-9','13600987654'),
                        (null,'刘燕红','女','1991-2-17','15960876767'),
                        (null,'杨鸣州','男','1990-12-27','13587112344'),
                        (null,'周茹平','女','1990-3-4','13098090988'),
                        (null,'林巧燕','女','1991-8-17','15265756543');
```

由于用户编号（id）设置为自动增长，所以 insert 命令中该字段不给出具体值。

#### 二、JDBC 实现数据库查询

查询操作是数据库最基本的操作，通过 select 语句对数据库进行查询，查询结果以 ResultSet 结果集的形式返回。以查询用户表中的所有用户信息为例，根据图 11-2-3 所示的 5 个步骤展开学习。

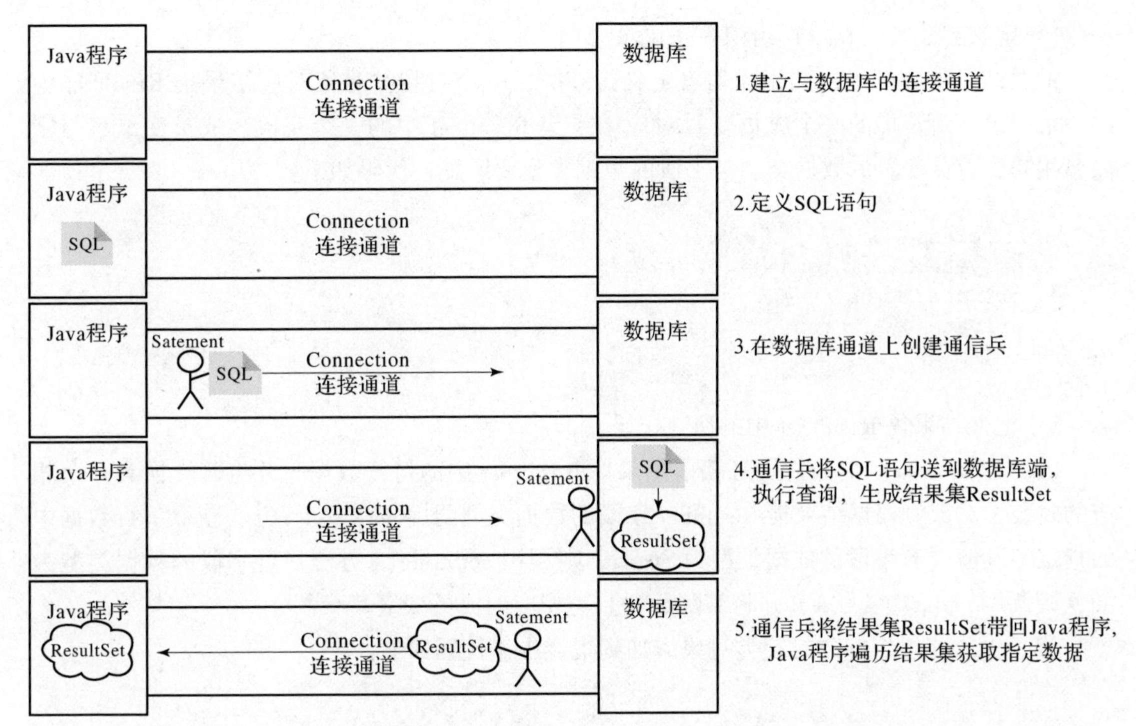


图 11-2-3　JDBC 实现数据库查询示意

### 1. 建立与数据库的连接通道

代码如下：

```
Class.forName("com.mysql.jdbc.Driver");
String URL = "jdbc:mysql://localhost:3306/javadb? useUnicode=true&
                                              characterEncoding=UTF-8";
String USER = "root";
String PWD = "123456";
Connection conn=DriverManager.getConnection(URL,USER,PWD);
```

### 2. 定义 SQL 语句

```
String sql="select * from User";
```

### 3. 在数据库通道上创建通信兵

```
Statement stmt=conn.createStatement();
```

### 4. 通信兵将 SQL 语句送到数据库端，执行查询，生成结果集 ResultSet

```
ResultSet rs=stmt.executeQuery(sql);
```

### 5. 通信兵将结果集 ResultSet 带回 Java 程序，Java 程序处理结果集获取指定数据

在通常情况下，结果集 ResultSet 包含数据库中的多条记录，Java 程序会以集合的形式组织这些记录。具体如下。

1）定义封装类（User）组织一条记录

根据结果集 ResultSet 的字段列表定义封装类，一个封装类对象对应结果集 ResultSet 中的一条记录，封装类的一个成员变量对应结果集 ResultSet 中的一个字段，成员变量名与字段名相同，成员变量的数据类型与字段的数据类型要匹配。代码如下：

```
public class User {
    private String id,name ,sex;
    private Date birthday;
    private String tel;
}
```

2）定义结果集 ResultSet 中的所有记录

（1）以 rs. next( ) 为条件循环获取结果集 ResultSet 中的每行数据，并将其转换为一个相应的封装类对象：根据结果集 ResultSet 字段的数据类型调用 getXXX( ) 方法获取一行数据中的指定字段值（若字段的数据类型为 String，则调用 getString( ) 方法；若字段的数据类型为 int，则调用 getInt( ) 方法），并将其赋给封装类相应的成员变量。

（2）循环将组织好的封装类对象添加到集合中。代码如下：

```
List<User> userList = new ArrayList<User>();
while(rs.next()) {//判断 rs 中是否还有未读取的数据行
    User user = new User();//定义空的封装类对象
    user.setId(rs.getString("id"));//从 rs 获取当前行 id 并赋给 user 对象
    user.setName(rs.getString("name"));
    user.setSex(rs.getString("sex"));
    user.setBirthday(rs.getDate("birthday"));
    user.setTel(rs.getString("tel"));
    userList.add(user);//将封装好的对象 user 添加到集合中
}
```

## 三、使用 PreparedStatement 接口实现数据库查询

PreparedStatement 接口继承 Statement 接口，用于执行预编译的 SQL 语句。此时 SQL 语句可以自带任意个参数，每个参数在 SQL 语句中用“?”作为占位符代替，具体参数值必须在 SQL 语句执行前通过 setter( ) 方法进行设定。

**【例 11-2-2】** 以查询 user 表中指定姓氏的用户为例，操作界面如图 11-2-4 所示。对比 Statement 接口与 PreparedStatement 接口的执行效率、安全性、代码可读性。

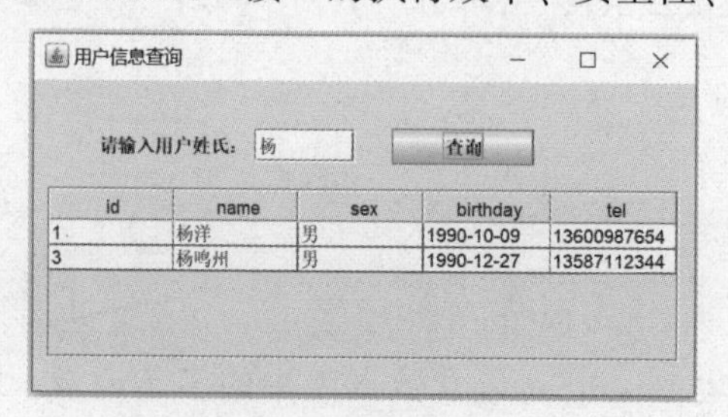


图 11-2-4　用户信息查询界面

代码如下：

```
1  JButton btnNewButton = new JButton("查询");
2      btnNewButton.addActionListener(new ActionListener() {
3          public void actionPerformed(ActionEvent e) {
4              //使用 Statement 接口
5              String search = textField.getText();
6              try {
7                  Class.forName("com.mysql.jdbc.Driver");
8                  String URL = "jdbc:mysql://localhost:3306/javadb?
                           useUnicode = true&characterEncoding = UTF - 8";
9                  String USER = "root";
10                 String PWD = "123456";
11                 Connection conn = DriverManager.getConnection(URL,USER,PWD);
12                 //使用 Statement 接口时,SQL 语句的定义
13                 String sql = "select * from user
                           where name like '" + search + "% '";
14                 Statement stmt = conn.createStatement();
15                 ResultSet rs = stmt.executeQuery(sql);
16                 //处理结果集
17                 Vector users = new Vector();
18                 while (rs.next()) {
19                     Vector row = new Vector();
20                     row.add(rs.getString("id"));
21                     row.add(rs.getString("name"));
22                     row.add(rs.getString("sex"));
23                     row.add(rs.getDate("birthday"));
24                     row.add(rs.getString("tel"));
25                     users.add(row);
26                 }
27                 Vector < String > columns = new Vector < String >();
28                   columns.add("id");
29                   columns.add("name");
30                  columns.add("sex");
31                  columns.add("birthday");
32                  columns.add("tel");
33                  //定义 JTable 数据模型
34                  TableModel model = new DefaultTableModel(users, columns);
35                  table.setModel(model);
36             } catch (Exception e1) {
37                  e1.printStackTrace();
38             }
39         }
40 });
```

【说明】

程序使用 Statement 接口作为执行对象时，第 13 行定义要执行的 SQL 语句，通过字符串的连接实现参数的引入。当存在多个参数时，SQL 语句的定义过于烦琐、可读性差，且每次执行时都必须重新编译，执行效率低。因此，引入 PreparedStatement 接口作为执行对象，对

Statement 接口进行改进，具体实现如下：

```
1  JButton btnNewButton = new JButton("查询");
2      btnNewButton.addActionListener(new ActionListener() {
3          public void actionPerformed(ActionEvent e) {
4              //使用 PrepareSstatement
5              String search = textField.getText();
6              try {
7                  Class.forName("com.mysql.jdbc.Driver");
8                  String URL = "jdbc:mysql://localhost:3306/javadb?
                           useUnicode = true&characterEncoding = UTF - 8";
9                  String USER = "root";
10                  String PWD = "123456";
11                  Connection conn = DriverManager.getConnection(URL,USER,PWD);
12                  //使用 PreparedStatement 接口时,SQL 语句的定义
13                  String sql = "select * from user where name like ?";
14                  //定义 PreparedStatement 接口对象
15                  PreparedStatement pstmt = conn.prepareStatement(sql);
16                  //PreparedStatement 接口对象执行前,必须设置 SQL 语句中"?"的数值
17                  pstmt.setString(1, search + "% ");
18                  ResultSet rs = pstmt.executeQuery();
19                  //处理结果集(同上)
20              } catch (Exception e1) {
21                  e1.printStackTrace();
22              }
23          }
24      });
```

【说明】

使用 PreparedStatement 接口作为执行对象，程序第 13 行 SQL 语句的参数使用“?”代替，简单易读，且如第 88 行参数单独传入，SQL 语句已进行了预编译，提高了执行效率，也防止了 SQL 注入的安全问题。值得注意的是，SQL 语句不作为 executeQuery( ) 方法的参数，而是作为创建 PreparedStatement 执行对象时的参数，如第 15 行。

### 11.2.3　更新数据库

#### 一、插入数据

使用 insert 语句将数据插入数据库表，必须注意参数的个数、顺序、数据类型是否与表中字段的数据类型匹配。执行更新操作（包括插入、修改、删除）的方法均为 executeUpdate( )，返回值为数据库执行 SQL 语句受影响的记录数。若返回值大于等于 1，则说明更新成功，否则更新失败。

**【例 11 – 2 – 3】** 向 user 表中添加一个新用户信息，操作界面如图 11 – 2 – 5 所示。

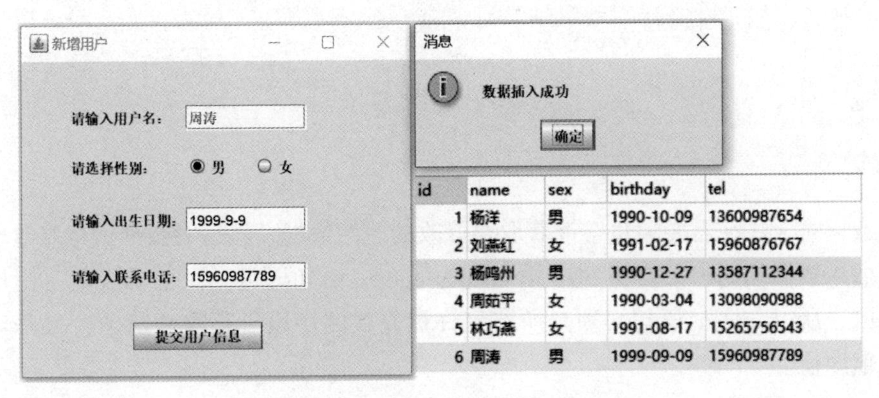


| id | name | sex | birthday | tel |
|---|---|---|---|---|
| 1 | 杨洋 | 男 | 1990-10-09 | 13600987654 |
| 2 | 刘燕红 | 女 | 1991-02-17 | 15960876767 |
| 3 | 杨鸣州 | 男 | 1990-12-27 | 13587112344 |
| 4 | 周茹平 | 女 | 1990-03-04 | 13098090988 |
| 5 | 林巧燕 | 女 | 1991-08-17 | 15265756543 |
| 6 | 周涛 | 男 | 1999-09-09 | 15960987789 |

图 11-2-5　添加用户信息界面

代码如下：

```
1  JButton btnNewButton = new JButton("提交用户信息");
2      btnNewButton.addActionListener(new ActionListener() {
3          public void actionPerformed(ActionEvent e) {
4              String name = textField.getText();
5             String sex;
6             if (rdbtnNewRadioButton.isSelected())
7                 sex = "男";
8             else
9                 sex = "女";
10             String birthday = textField_1.getText();
11             String tel = textField_2.getText();
12             try {
13                 Class.forName("com.mysql.jdbc.Driver");
14                 String URL = "jdbc:mysql://localhost:3306/javadb?
                                useUnicode=true&characterEncoding=UTF-8";
15             String USER = "root";
16             String PWD = "123456";
17             Connection conn = DriverManager.getConnection(URL,USER,PWD);
18             //使用 PreparedStatement
19             String sql = "insert into user values(null,?,?,?,?)";
20             PreparedStatement pstmt = conn.prepareStatement(sql);
21             //PreparedStatement 接口对象执行前,必须设置 SQL 语句中"?"的数值
22             pstmt.setString(1, name);
23             pstmt.setString(2, sex);
24             pstmt.setString(3, birthday);
25             pstmt.setString(4, tel);
26             int n = pstmt.executeUpdate();
27             if (n >=1)
28                 JOptionPane.showMessageDialog(null ,"数据插入成功");
29             else
30                 JOptionPane.showMessageDialog(null ,"数据插入失败");
31         } catch (Exception e1) {
32             e1.printStackTrace();
33             }
```

```
34          }
35  });
```

## 二、修改数据

修改数据时，往往需要通过查询获取要修改的原始数据，然后对数据进行修改。具体实现过程与插入数据相同，只要将 SQL 语句改为 update 语句即可。由于无法确定用户具体修改哪些字段，所以 update 语句必须对所有允许被修改的字段进行重新赋值，只是对未被修改的字段赋原值。

**【例 11－2－4】** 修改 user 表中杨洋的联系电话，操作界面如图 11－2－6 所示。

（1）查询要修改的原始数据。

（2）修改数据。

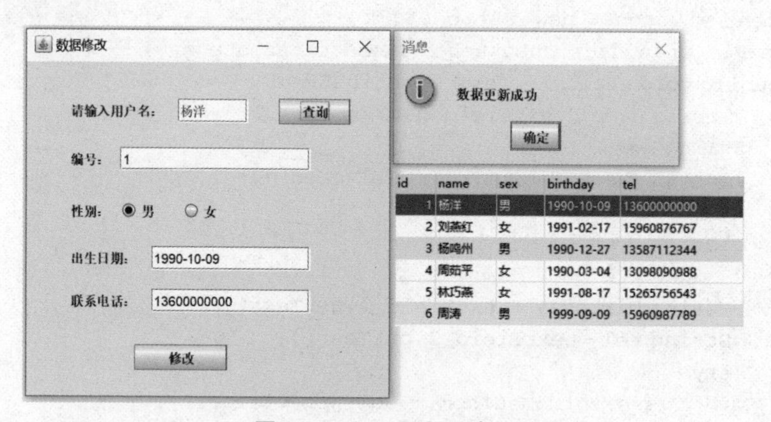


图 11－2－6　更新用户信息界面

代码如下：

```
1  JButton btnNewButton_1 = new JButton("修改");
2      btnNewButton_1.addActionListener(new ActionListener() {
3          public void actionPerformed(ActionEvent e) {
4              String name = textField.getText();
5              String sex;
6              if (rdbtnNewRadioButton.isSelected())
7              sex = "男";
8          else
9              sex = "女";
10          String birthday = textField_2.getText();
11          String tel = textField_3.getText();
12          try {
13              Class.forName("com.mysql.jdbc.Driver");
14              String URL = "jdbc:mysql://localhost:3306/javadb?
                        useUnicode = true&characterEncoding = UTF - 8";
15              String USER = "root";
16              String PWD = "123456";
```

```
17              Connection conn = DriverManager.getConnection(URL,USER,PWD);
18              String sql = "update user set sex = ?,birthday = ?,tel = ?
                        where name = ?";
19              PreparedStatement pstmt = conn.prepareStatement(sql);
20              pstmt.setString(1, sex);
21              pstmt.setString(2, birthday);
22              pstmt.setString(3, tel);
23              pstmt.setString(4, name);
24              int n = pstmt.executeUpdate();
25              if (n > =1)
26                  JOptionPane.showMessageDialog(null,"数据更新成功");
27              else
28                  JOptionPane.showMessageDialog(null,"数据更新失败");
29          } catch (Exception e1) {
30                  e1.printStackTrace();
31              }
32          }
33  });
```

## 三、删除数据

删除数据使用 delete 语句。

**【例 11－2－5】** 根据用户姓名删除用户信息，操作界面如图 11－2－7 所示。

图 11－2－7 删除用户信息界面

代码如下：

```
1  JButton btnNewButton  = new JButton("删除");
2      btnNewButton.addActionListener(new ActionListener() {
3          public void actionPerformed(ActionEvent e) {
4              String name = textField.getText();
5              try {
6                  Class.forName("com.mysql.jdbc.Driver");
7                  String URL  = "jdbc:mysql://localhost:3306/javadb?
                           useUnicode = true&characterEncoding = UTF - 8";
8                  String USER  = "root";
9                  String PWD  = "123456";
10                 Connection conn = DriverManager.getConnection(URL,USER,PWD);
11                 String sql = "delete from user where name = ?";
12                 PreparedStatement pstmt = conn.prepareStatement(sql);
13                 pstmt.setString(1, name);
14                 int n = pstmt.executeUpdate();
15                 if (n >=1)
```

```
16                  JOptionPane.showMessageDialog(null,"数据删除成功");
17              else
18                  JOptionPane.showMessageDialog(null,"数据删除失败");
19          } catch (Exception e1) {
20              e1.printStackTrace();
21          }
22      }
23  });
```

### 11.2.4 关闭与数据库的连接

JDBC 访问数据库，创建了 Connection 接口对象、Statement 接口对象或者 PreparedStatement 接口对象、ResultSet 接口对象，这些对象占用了一定的数据库资源及网络资源，因此，每次结束对数据库的访问后，都要调用 close( ) 方法及时销毁这些对象，释放相应资源。为了保证销毁代码一定会被执行，将销毁代码放在异常捕获语句的 finally 部分。建议先创建的对象后销毁，后创建的对象先销毁。

以【例 11-2-2】为例，一个使用 JDBC 访问数据库的完整程序如下：

```
try {
    Class.forName("com.mysql.jdbc.Driver");
    String URL = "jdbc:mysql://localhost:3306/javadb?
                useUnicode=true&characterEncoding=UTF-8";
    String USERNAME = "root";
    String PASSWORD = "123456";
    conn=DriverManager.getConnection(URL,USERNAME,PASSWORD);
    String sql="select * from user where name like ?";
    pstmt=conn.prepareStatement(sql);
    pstmt.setString(1, search+"%");
    rs=pstmt.executeQuery();
    //处理结果集(略)
} catch (Exception e1) {
    e1.printStackTrace();
}finally {
    try {
        if (rs!=null)rs.close();
        if (pstmt!=null)pstmt.close();
        if (conn!=null)conn.close();
    } catch (SQLException e1) {
        e1.printStackTrace();
    }
 }
```

**任务工单**

<table>
<tr><td>任务名称</td><td colspan="5">图书信息管理系统——从数据库查询所有图书信息</td></tr>
<tr><td>班级</td><td></td><td>学号</td><td></td><td>姓名</td><td></td></tr>
<tr><td>任务要求</td><td colspan="5">(1) 打开 Navicat，导入数据库 bookManagement。<br>(2) 根据 book 表（下图），定义封装类 Book。<br>book<br>bookID: char(3)<br>bookName: char(30)<br>writer: char(20)<br>press: char(10)<br>price: double<br>status: char(2)<br>(3) 创建测试类 Test，从数据库查询所有图书信息并输出。</td></tr>
<tr><td>任务实现</td><td colspan="5"><pre>public class Test {
        public static void main(String[ ] args) {



        }
    }</pre></td></tr>
<tr><td>异常及待解决问题记录</td><td colspan="5"></td></tr>
</table>

## 11.3 MVC 模式应用开发

### 学习目标

（1）了解什么是 MVC 模式。

（2）掌握 MVC 应用开发方法。

### 相关知识

MVC 模式即 Model－View－Controller（模型－视图－控制器）模式，该模式用于应用程序的分层开发。传统 MVC 模式分为以下 3 层。

（1）Model（模型）：表示企业数据和业务规则，可以是一个封装类。

（2）View（视图）：用户与应用程序进行交互的界面，可以是一个 GUI。

（3）Controller（控制器）：作用于模型和视图上，控制数据流向模型对象，并在数据变化时更新视图，使视图与模型分离。

11.2 节通过 JDBC 实现了对数据库中用户表的增、删、改、查功能，从程序中不难发现存在大量重复书写的代码，例如数据库的连接代码。为了提高代码的重用性，降低代码的耦合度，提高执行效率，本节介绍如何使用 MVC 模式进行项目开发。

### 一、项目分层及其初始化

基于 MVC 模式，将项目分为以下 5 层。

（1）bean 层（或称为 model 层）：数据实体模型层、实体层，存放实体类。

（2）view 层：视图层，负责界面的交互。

（3）service 层：业务层，处理所有业务逻辑，调用 dao 层接口。

（4）dao 层：数据访问层，进行数据业务处理和持久化操作。

（5）utils 层：工具层，用于存放工具类。

项目分层初始的过程如下。

（1）创建项目：新建 Java 项目。

（2）创建 5 层结构：在 src 目录下创建 5 个包，分别为 bean、view、service、dao、utils。

①utils 层：添加 DBUtil 类。

DBUtil 类由静态代码块加载数据库驱动程序，由成员方法 getConnection( ) 负责建立与数据库的连接，由成员方法 close( ) 实现断开与数据库的连接。代码如下：

```
1  public class DBUtil {
2      private static final String DRIVER = "com.mysql.jdbc.Driver";
```

```
3       private static final String URL = "jdbc:mysql://localhost:3306/javadb?
                                          characterEncoding=UTF-8";
4       private static final String USER = "root";
5       private static final String PWD = "123456";
6       private static Connection conn = null;
7    private static ThreadLocal<Connection> tl = new ThreadLocal<Connection>();
8       //使用静态代码块加载驱动程序
9       static {
10          try {
11              Class.forName(DRIVER);
12          } catch (ClassNotFoundException e) {
13              e.printStackTrace();
14      }
15      }
16      //定义一个获取数据库连接的方法
17      public static Connection getConnection() {
18          try {
19              conn = DriverManager.getConnection(URL, USER, PwD);
20          } catch (SQLException e) {
21              e.printStackTrace();
22          }
23          return conn;
24      }
25      //关闭回收数据源
26   public static void close(ResultSet rs, Statement stat, Connection conn) {
27          try {
28              if(rs != null)   rs.close();
29              if(stat != null)stat.close();
30              if(conn != null)conn.close();
31          } catch (SQLException e) {
32              e.printStackTrace();
33          }
34      }
35  }
```

②bean 层：参考数据库定义封装类，数据库中的 char 数据类型可定义为 String 数据类型，date 数据类型可定义为 java. util. Date 数据类型，bit 数据类型可定义为 boolean 数据类型。

③service 层：创建 XXXService 类，实现业务逻辑。

④dao 层：创建 XXXDao 类，封装对表的增、删、改、查等操作。

## 二、各层间的调用及其参数传递

以【例 11-2-2】实现用户信息管理系统的条件查询功能为例，各层间的调用及其参数传递如图 11-3-1 所示。

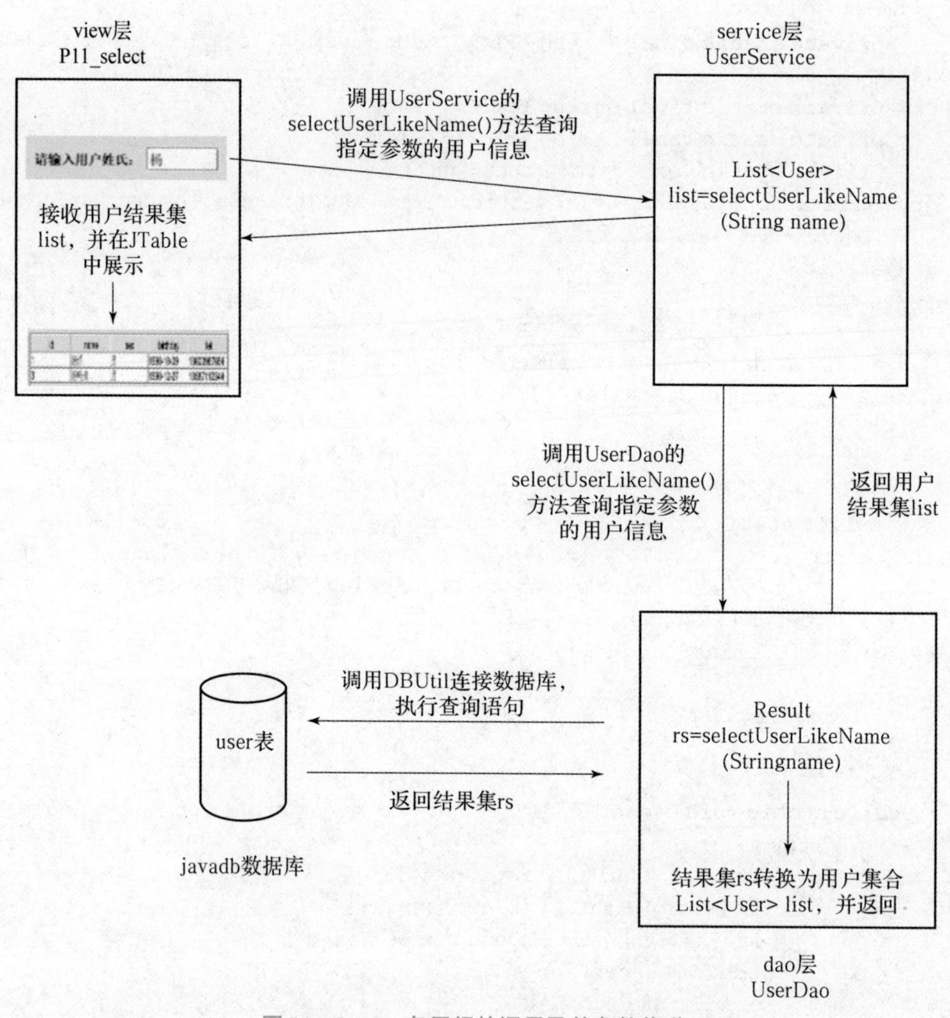


图 11－3－1　各层间的调用及其参数传递

**1. 各层间的调用**

view 层调用 service 层的查询方法 selectUserLikeName（String name），servcie 层调用 dao 层的查询方法 selectUserLikeName（String name），dao 层的 selectUserLikeName（String name）方法调用 utils 层的 DBUtil 实现数据库的连接，并实现数据的查询。

**2. 参数传递**

view 层获取参数 String name，传入 service 层的查询方法 selectUserLikeName（String name），最后传入 dao 层的查询方法 selectUserLikeName（String name）。

**3. 返回值传递**

dao 层的查询方法 selectUserLikeName（String name）执行查询，获取结果集 rs，将 rs 转换为用户集合 List < User > list，并传回 service 层；service 层也将返回的用户集合 List < User > list 传回 view 层；view 层则以 JTable 的形式将用户集合 List < User > list 展示给用户。

## 三、功能实现

### 1. view 层（SelectUser 窗体）

代码如下：

```
19 JButton btnNewButton = new JButton("查询");
20     btnNewButton.addActionListener(new ActionListener() {
21         public void actionPerformed(ActionEvent e) {
22             String search = textField.getText();
23             //调用 service 层的 selectUserLikeName()方法获取查询结果
24             UserService userService = new UserService();
25             List < User > list = userService.selectUserLikeName(search);
26             //定义 JTable 数据模型
27             Vector users = new Vector();
28             for (User user : list) {
29                 Vector row = new Vector();
30                 row.add(user.getId());
31                 row.add(user.getName());
32                 row.add(user.getSex());
33                 row.add(user.getBirthday());
34                 row.add(user.getTel());
35                 users.add(row);
36             }
37             Vector < String > columns = new Vector < String >();
38               columns.add("id");
39               columns.add("name");
40               columns.add("sex");
41               columns.add("birthday");
42               columns.add("tel");
43         DefaultTableModel model = new DefaultTableModel(users,columns);
44               table.setModel(model);
45     }
46 });
```

### 2. service 层（UserService）

代码如下：

```
2  public class UserService {
3      UserDao userDao;
4      public UserService() {
5          userDao = new UserDao();
6      }
7      public List < User > selectUserLikeName(String search) {
8          List < User > list = userDao.selectUserLikeName(search);
9          return list;
10     }
11 }
```

### 3. dao 层（UserDao）

代码如下：

```
2  public class UserDao {
3      private Connection conn;
4      private PreparedStatement pstmt;
5      private ResultSet rs;
6  
7      public List <User > selectUserLikeName(String search) {
8          List <User > list = new ArrayList <User >();
9          try {
10             //调用 DBUtil 的 getCconnection()方法获取与数据库的连接通道
11             conn = DBUtil.getConnection();
12             //使用 PreparedStatement 接口时,SQL 语句的定义
13             String sql = "select * from user where name like ?";
14             //定义 PreparedStatement 接口对象
15             pstmt = conn.prepareStatement(sql);
16             //PreparedStatement 接口对象执行前,必须设置 SQL 语句中"?"的值
17             pstmt.setString(1, search + "%");
18             rs = pstmt.executeQuery();
19             while (rs.next()) {
20                 User u = new User();
21                 u.setId(rs.getString("id"));
22                 u.setName(rs.getString("name"));
23                 u.setSex(rs.getString("sex"));
24                 u.setBirthday(rs.getDate("birthday"));
25                 u.setTel(rs.getString("tel"));
26                 list.add(u);
27             }
28         } catch (Exception e) {
29             e.printStackTrace();
30         }finally {
31             DBUtil.close(rs, pstmt, conn);
32         }
33         return list;
34     }
35 }
```

### 案例

使用 MVC 模式实现用户数据的查询、增加、删除、修改功能。

### 1. bean 层

代码如下：

```
public class User {
    private String id,name ,sex;
    private Date birthday;
    private String tel;
}
```

**2. view 层**

新增用户信息界面如图 11-3-2 所示，代码如下：

图 11-3-2　新增用户信息界面

```
JButton btnNewButton  = new JButton("提交用户信息");
    btnNewButton.addActionListener(new ActionListener() {
        public void actionPerformed(ActionEvent e) {
            String name = textField.getText();
            String sex;
            if (rdbtnNewRadioButton.isSelected())
                sex = "男";
        else
                sex = "女";
            String birthday = textField_1.getText();
            String tel = textField_2.getText();
            User u = new User();
            u.setName(name);
            u.setSex(sex);
            SimpleDateFormat format = new SimpleDateFormat("yyyy-MM-dd");
            Date birth = null ;
            try {
                birth  = format.parse(birthday);
            } catch (ParseException e1) {
                e1.printStackTrace();
            }
            u.setBirthday(birth);
            u.setTel(tel);
            UserService userService = new UserService();
            int n = userService.insertUser(u);
            if (n >=1)
                 JOptionPane.showMessageDialog(null ,"数据插入成功");
            else
                 JOptionPane.showMessageDialog(null ,"数据插入失败");
        }
  });
```

修改用户信息界面如图 11－3－3 所示，代码如下：

图 11－3－3　修改用户信息界面

```
JButton btnNewButton_1 = new JButton("修改");
    btnNewButton_1.addActionListener(new ActionListener() {
        public void actionPerformed(ActionEvent e) {
            String name = textField.getText();
            String sex;
            if (rdbtnNewRadioButton.isSelected())
                sex = "男";
            else
                sex = "女";
            String birthday = textField_2.getText();
            String tel = textField_3.getText();
            User u = new User();
            u.setName(name);
            u.setSex(sex);
            SimpleDateFormat format = new SimpleDateFormat("yyyy-MM-dd");
            Date birth = null ;
            try {
                birth = format.parse(birthday);
            } catch (ParseException e1) {
                e1.printStackTrace();
            }
            u.setBirthday(birth);
            u.setTel(tel);
            UserService userService = new UserService();
            int n = userService.updateUserByName(u);
            if (n >= 1)
                JOptionPane.showMessageDialog(null ,"数据更新成功");
            else
                JOptionPane.showMessageDialog(null ,"数据更新失败");
        }
    });
```

删除用户信息界面如图 11－3－4 所示，代码如下：

图 11－3－4　删除用户信息界面

```
JButton btnNewButton = new JButton("删除");
    btnNewButton.addActionListener(new ActionListener() {
        public void actionPerformed(ActionEvent e) {
            String name = textField.getText();
            UserService userService = new UserService();
            int n = userService.deleteUserByName(name);
            if (n > =1)
                JOptionPane.showMessageDialog(null,"数据删除成功");
            else
                JOptionPane.showMessageDialog(null,"数据删除失败");
        }
    });
```

### 3. service 层

代码如下：

```
public class UserService {
    UserDao userDao;
    public UserService() {
        userDao = new UserDao();
    }
    public List<User> selectUserLikeName(String search) {
        return userDao.selectUserLikeName(search);
    }
    public int insertUser(User u) {
        return userDao.insertUser(u);
    }
    public int updateUserByName(User u) {
        return userDao.updateUserByName(u);
    }
    public int deleteUserByName(String name) {
        return userDao.deleteUserByName(name);
    }
}
```

### 4. dao 层

代码如下：

```
public class UserDao {
    private Connection conn;
    private PreparedStatement pstmt;
    private ResultSet rs;
    public List<User> selectUserLikeName(String search) {
        List<User> list = new ArrayList<User>();
        try {
            conn = DBUtil.getConnection();
            String sql = "select * from user where name like ?";
            pstmt = conn.prepareStatement(sql);
            pstmt.setString(1, search + "%");
            rs = pstmt.executeQuery();
            while (rs.next()) {
                User u = new User();
                u.setId(rs.getString("id"));
                u.setName(rs.getString("name"));
                u.setSex(rs.getString("sex"));
                u.setBirthday(rs.getDate("birthday"));
                u.setTel(rs.getString("tel"));
                list.add(u);
            }
        } catch (Exception e) {
            e.printStackTrace();
        }finally {
            DBUtil.close(rs, pstmt, conn);
        }
        return list;
    }
    public int insertUser(User u) {
        int n = 0;
        try {
            conn = DBUtil.getConnection();
            String sql = "insert into user values(null,?,?,?,?)";
            PreparedStatement pstmt = conn.prepareStatement(sql);
            pstmt.setString(1, u.getName());
            pstmt.setString(2, u.getSex());
            pstmt.setDate(3, new java.sql.Date(u.getBirthday().getTime()));
            pstmt.setString(4, u.getTel());
            n = pstmt.executeUpdate();
        } catch (Exception e) {
            e.printStackTrace();
        }finally {
            DBUtil.close(rs, pstmt, conn);
        }
        return n;
    }
    public int updateUserByName(User u) {
    int n = 0;
    try {
        conn = DBUtil.getConnection();
        String sql = "update user set sex = ?,birthday = ?,tel = ? where
                                                    name = ?";
```

```
            PreparedStatement pstmt = conn.prepareStatement(sql);
            pstmt.setString(1, u.getSex());
            pstmt.setDate(2, new
                            java.sql.Date(u.getBirthday().getTime()));
            pstmt.setString(3, u.getTel());
            pstmt.setString(4, u.getName());
            n = pstmt.executeUpdate();
        } catch (Exception e) {
            e.printStackTrace();
        }finally {
            DBUtil.close(rs, pstmt, conn);
        }
        return n;
    }
    public int deleteUserByName(String name) {
        int n = 0;
        try {
            conn = DBUtil.getConnection();
            String sql = "delete from user where name = ?";
            PreparedStatement pstmt = conn.prepareStatement(sql);
            pstmt.setString(1, name);
            n = pstmt.executeUpdate();
        } catch (Exception e) {
            e.printStackTrace();
        }finally {
            DBUtil.close(rs, pstmt, conn);
        }
        return n;
    }
}
```

## 任务工单

<table>
<tr><td>任务名称</td><td colspan="5">图书信息管理系统——从数据库查询所有图书信息</td></tr>
<tr><td>班级</td><td></td><td>学号</td><td></td><td>姓名</td><td></td></tr>
<tr><td>任务要求</td><td colspan="5">（1）打开 Navicat，导入数据库 bookManagement。<br>（2）根据 book 表（下图），定义封装类 Book。<br>book<br>bookID: char(3)<br>bookName: char(30)<br>writer: char(20)<br>press: char(10)<br>price: double<br>status: char(2)<br>（3）使用 MVC 模式，从数据库查询所有图书信息，以表格展示结果。</td></tr>
<tr><td>任务实现</td><td colspan="5">（1）util 层：<br>（2）bean 层：<br>（3）view 层：<br>（4）service 层：<br>（5）dao 层：</td></tr>
<tr><td>异常及待解决问题记录</td><td colspan="5"></td></tr>
</table>

## 练习题

### 1. 选择题

(1) 下列关于 JDBC 的说法正确的是（　　）。

A. 使用 JDBC 连接数据库，必须使用连接池

B. 使用 JDBC 连接数据库，若使用连接池就不需要导入数据库驱动程序

C. 操作不同的数据库可以使用相同的驱动程序

D. 无论如何，只要使用 JDBC，那么就要使用 JDBC 驱动程序

(2) DriverManager 类的 getConnection( ) 方法的作用是（　　）。

A. 取得数据库连接　　B. 取得数据表

C. 取得字段　　D. 取得记录

(3) Class 类的 forName( ) 方法的作用是（　　）。

A. 注册类名　　B. 注册数据库驱动程序

C. 创建类名　　D. 创建数据库驱动程序

(4) Connection 类的 createStatement( ) 方法的作用是（　　）。

A. 创建数据库　　B. 创建数据表

C. 创建记录集　　D. 创建 SQL 命令执行接口

(5) ResultSet 接口的 next( ) 方法的作用是（　　）。

A. 取得下一条记录　　B. 取得下两条记录

C. 取得上两条记录　　D. 取得上一条记录

(6) 下列不属于 JDBC 基本功能的是（　　）。

A. 与数据库建立连接　　B. 提交 SQL 语句

C. 处理查询结果　　D. 维护管理数据库

(7) 下列关于 ResultSet 接口中游标指向的描述正确的是（　　）。

A. ResultSet 接口对象初始化时，游标在表格的第一行

B. ResultSet 接口对象初始化时，游标在表格的第一行之前

C. ResultSet 接口对象初始化时，游标在表格的最后一行之前

D. ResultSet 接口对象初始化时，游标在表格的最后一行

(8) 下列关于 Statement 接口的描述错误的是（　　）。

A. Statement 接口用于执行 SQL 语句的

B. Statement 接口是 PreparedStatement 接口的子接口

C. 获取 Statement 接口实例需要使用 Connection 接口的 createStatement( ) 方法

D. PreparedStatement 接口能使用参数占位符，而 Statement 接口不行

### 2. 填空题

(1) __________是一种用于执行 SQL 语句的 Java API，提供了访问和操作关系数据库的方法。

（2）JDBC 驱动程序按照工作方式分为 4 种：__________ 和__________、__________、__________。

（3）JDBC 加载指定驱动程序后，就可以利用 DriveManager 类的静态方法建立与________的连接。

（4）通过连接对象创建执行对象，实现对数据库的插入、修改、删除和查询操作。常见的执行对象有 3 种类型：________、__________、__________。

（5）ResultSet 接口中定义了大量的 getXXX( ) 方法，如果使用字段的索引获取指定的数据，则字段的索引从________开始。

第 12 章

# 网络编程

——书山有路，学海无涯

# 12.1 网络编程基础

## 学习目标

（1）了解 TCP/IP。

（2）掌握 InetAddress 类的使用方法。

## 相关知识

计算机网络将处于不同地理位置的计算机连接起来，同一个网络中的计算机必须遵守一定的规则才能实现相互访问与通信，这些规则就是网络通信协议，它规定了主机间的寻址规则、数据的传输格式及传输方式等。目前使用最广泛的网络协议是传输控制协议/因特网互连协议（Transmission Control Protocol/Internet Protocol，TCP/IP）。TCP/IP 分为 4 层，分别是链路层、网络层、传输层、应用层，每层负责不同的通信功能，下层为上层提供服务，如图 12－1－1 所示。

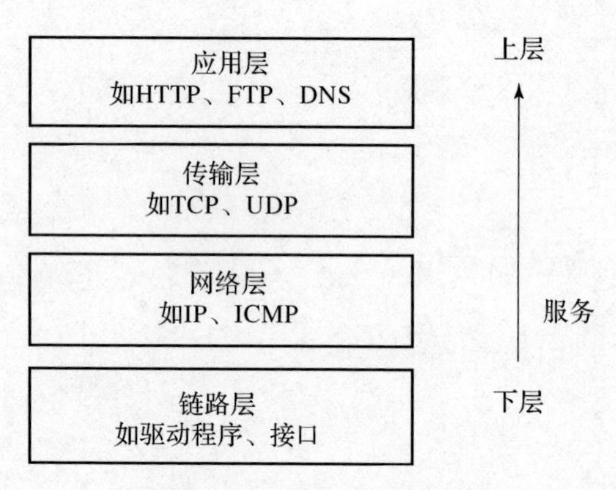


图 12－1－1　TCP/IP 分层

（1）链路层：负责监视数据在主机和网络之间的交换。

（2）网络层：负责路由以及把分组报文发送给目标网络或主机。

（3）传输层：负责对报文进行分组和重组，并以 TCP 或 UDP（User Datagram Protocol）格式封装报文。TCP 提供了可靠的面向连接的数据传输服务，而 UDP 提供了不可靠的面向非连接的数据传输服务。

（4）应用层：负责向用户提供应用程序。

## 一、IP 地址

在网络通信中，每台主机都必须有一个唯一的标识号，通过这个标识号确定主机的定位，实现数据传输的路由选择。在 TCP/IP 中，这个标识号就是 IP 地址，它可以唯一标识一台主机。

IP 地址由两部分组成，即“网络地址．主机地址”。其中，网络地址表示其属于互联网中的哪个网络，主机地址表示其属于该网络中的哪台主机。IP 地址广泛使用的版本是 IPv4，它由 4 字节的二进制数表示，但由于二进制形式的 IP 地址不便于记忆，所以 IP 地址通常写成十进制形式，每个字节用 1 个十进制数字（0 ~ 255）表示，数字间用“.”分隔，例如“127. 0. 0. 1”指本机 IP 地址，通常用于测试。

## 二、端口号

通过 IP 地址可以连接到指定计算机，但如果想访问目标计算机中的某个应用程序，还需要指定端口号。在计算机中，不同的应用程序有各自的端口号，例如 MySQL 数据库所使用的端口号为 3306。端口号是用 2 个字节表示的，它的取值范围是 0 ~ 65 535。其中，0 ~ 1 023 的端口号由操作系统的网络服务占用，普通应用程序需要使用 1 024 以上的端口号，从而避免端口号被其他应用程序或服务占用。下面通过图 12 – 1 – 2 解析 IP 地址和端口号的作用。

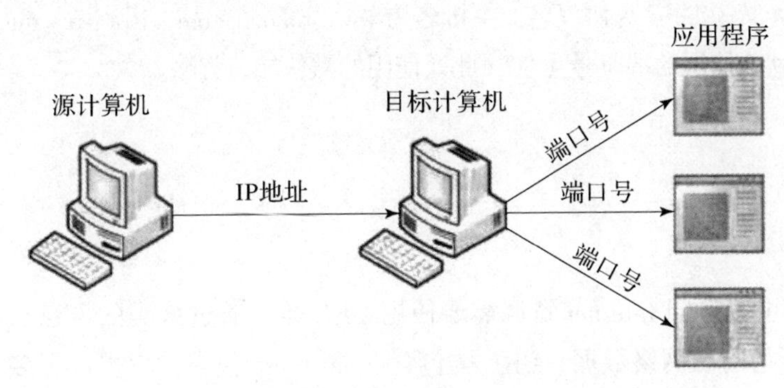


图 12 – 1 – 2　IP 地址和端口号的作用示意

## 三、TCP 与 UDP

TCP 和 UDP 属于传输层协议。

（1）TCP 称为传输控制协议，是一个面向连接的通信协议。也就是说，在传输数据前必须建立发送端和接收端的逻辑连接。TCP 提供了两台计算机之间的可靠无差错的数据传输。在 TCP 连接中必须明确客户端与服务器端，由客户端向服务器端发出连接请求，每次连接的创建都需要经过“3 次握手”。第 1 次握手，客户端向服务器端发出连接请求，等待服务器端确认；第 2 次握手，服务器端向客户端回送一个响应，通知客户端收到了连接请求；第 3 次握手，客户端再次向服务器端发送确认信息，确认连接。

（2）UDP 称为用户数据报协议，是无连接通信协议，即在数据传输时，数据的发送端和接收端不建立逻辑连接。也就是说，一台计算机向另外一台计算机发送数据时，发送端不管接收端是否存在，都直接发送数据，而接收端收到数据后，也不会向发送端回送反馈信息。不难发现，UDP 适用于可靠性要求较低、注重传输效率的应用，例如音频、视频数据的传输，偶尔丢失一两个数据包也不会对接收结果产生太大影响。

# 12.2 URL

## 学习目标

（1）理解 URL 的作用。

（2）掌握 URL 类的使用方法。

（3）掌握 URLConnection 类的使用方法。

## 相关知识

Internet 中的所有网络资源都用 URL 进行定位，一个 URL 通常由 4 个部分组成，分别为协议名、主机名、资源路径标识、端口号。例如，访问 JSP 菜鸟教程 https：// www. runoob. com/jsp/jsp - tutorial. html，其中 https 为协议名，主机名为 www. runoob. com，jsp/jsp - tutorial. html 为资源路径标识，https 默认使用的端口号为 80，此处使用的默认端口省略。

### 12.2.1 URL 类

#### 一、URL 类的构造方法

在 Java 程序中，访问 Internet 资源需要使用 URL 类，通过该类的方法获取网络数据流，读取来自 URL 所指向的网络数据。URL 类包含在 java. net 包中，它提供了多个构造方法。

(1)public URL(String spec)。
(2)public URL(URL context, String spec)。
(3)public URL(String protocol, String host, int port, String file)。
(4)public URL(String protocol, String host, String file)。

例如：

```
URL url1 = new URL("https://www.sina.com");
URL url2 = new URL(url1,"index.html");
URL url3 = new URL("https","www.sina.com","index.html");
URL url4 = new URL("https","www.sina.com",80,"index.html");
```

#### 二、URL 类的常用方法

URL 类的常用方法如表 12 - 2 - 1 所示。

表 12 - 2 - 1　URL 类的常用方法

| 方法 | 功能描述 |
|---|---|
| String getProtocol() | 获取该 URL 的协议名 |

续表

| 方法 | 功能描述 |
| --- | --- |
| String getHost( ) | 获取该 URL 的主机名 |
| int getPort( ) | 获取 URL 的端口号，若没有则返回 -1 |
| String getFile( ) | 获取 URL 的文件名，若没有则返回空串 |
| URLConnection openConnection( ) | 获取与 URL 进行连接的 URLConnection 类对象 |
| InputStream openStream( ) | 打开与 URL 的连接，返回来自连接的输入流 |
| Object getContent( ) | 获取 URL 的内容 |

**【例 12-2-1】** 读取新浪网主页内容。代码如下：

```
1  public class P12_2_1 {
2      public static void main(String[] args) {
3          try {
4              URL url = new URL("https://www.sina.com");
5              BufferedReader br = new BufferedReader(new
                                       InputStreamReader(url.openStream()));
6              String s = null;
7             while ((s = br.readLine()) != null)
8                 System.out.println(s);
9             br.close();
10         } catch (Exception e) {
11             //TODO Auto-generated catch block
12             e.printStackTrace();
13          }
14     }
15  }
```

运行结果如图 12-2-1 所示。

```
<!DOCTYPE html>
<html>
<head>
        <title>home.sina.com</title>
        <meta charset="utf-8">
        <meta name="viewport" content="width=device-width, initial-scale=1" >

        <link rel="stylesheet" type="text/css" href="https://image2.sina.com.cn/img/beimei_style.css">
        <style>
                .footer{width:100%;overflow:hidden;text-align:center;color:#939393;font-size:12px}.foote
        </style>
</head>
<body><!-- body code begin -->
```

图 12-2-1　运行结果

【说明】

程序第 4 行确定要访问的网络资源为新浪网主页，第 5 行通过调用 openStream( ) 方法获取读取新浪网主页内容的输入流，第 13、14 行循环读取新浪网主页内容，并在控制台输出，第 9 行关闭输入流。输出结果为新浪网主页的源码。

## 12.2.2　URLConnection 类

使用 URL 类可以读取 URL 指定的资源，如果要与 URL 指定的资源进行双向通信，则必须使用 URLConnection 类。通过 URL 类的 openConnection( )方法，URLConnection 类将创建一个与 URL 指定的资源连接对象，通过这个 URLConnection 类对象创建输入流、输出流实现双向通信。URLConnection 类的常用方法如表 12－2－2 所示。

表 12－2－2　URLConnection 类的常用方法

| 方法 | 功能描述 |
| --- | --- |
| void connect( ) | 打开 URL 指定资源的通信链路 |
| int getContentLength( ) | 返回 URL 的内容长度 |
| InputStream getInputStream( ) | 返回来自连接的输入流 |
| OutputStream getOutputStream( ) | 返回写往连接的输出流 |

**【例 12－2－2】** 代码如下：

```
1  public class P12_2_2 {
2      public static void main(String[] args) {
3          try {
4              URL url = new URL("http://101.34.44.3:8086/v1/customer/register");
5              URLConnection conn = url.openConnection();
6              String custStr = "{\"name\":\"Tom\",\"pwd\":\"123\"}";
7              System.out.println("custStr:" + custStr);
8              OutputStream os = conn.getOutputStream();
9              OutputStreamWriter osw = new OutputStreamWriter(os);
10             BufferedWriter bw = new BufferedWriter(osw);
11             bw.write(custStr);
12             bw.flush();
13             InputStream is = conn.getInputStream();
14             InputStreamReader isr = new InputStreamReader(is);
15             BufferedReader br = new BufferedReader(isr);
16             String line = "";
17             while ((line = br.readLine()) != null) {
18                 System.out.println(line);
19             }
20         } catch (Exception e1) {
21             e1.printStackTrace();
22         }
23     }
24 }
```

【说明】

程序第 4 行定义了要访问的资源为 101.34.44.3 服务器上的网站 v1 中定义的 register 接口；第 5 行创建了与 register 接口进行双向通信的 URLConnection 类对象 conn；第 6 行定义要写入 register 接口的字符串；第 8 行调用 URLConnection 类的 getOutputStream( )方法获取写往 register 接口的输出流；第 11、12 行将字符串写入 register 接口；第 13 行调用 URLConnection 类的 getInputStream( )方法获取来自 register 接口的输入流；第 17～19 行循环读取 register 接口返回值的每行数据并输出。

# 12.3 Socket通信

## 学习目标

（1）了解Socket编程原理。

（2）掌握Socket类和ServerSocket类的使用方法。

## 相关知识

在网络通信中，当应用层通过传输层进行数据通信时，会出现多个TCP连接或者多个应用程序进程需要通过通一个TCP端口传输数据的情况，为了区分不同的应用程序进程和连接，许多计算机操作系统会为应用程序与TCP/IP提供Socket接口。Socket通常用来实现客户端与服务器端的连接。在Java中，一个Socket由一个IP地址和一个端口号唯一确定，通过这个Socket接口可以与任何一台具有Socket接口的计算机通信。

下面介绍Java的Socket编程原理。

java.net包中定义了两个类：Socket类和ServerSocket类。它们分别用于Socket通信的客户端和服务器端，通过这两个类可以实现Internet中任意两台计算机的Socket通信。ServerSocket所在的服务器端每次只允许一个客户端与之连接，如果存在多个客户端请求连接，则这些请求被存入队列，队列大小默认为50。

## 一、Socket类

### 1. Socket类的构造方法

Socket类在客户端使用，通过构造一个Socket类对象与服务器端建立连接，Socket连接既可以是流连接，也可以是数据包连接，一般选择流连接，其数据传送准确有序，但速度较低。Socket类的常用构造方法如下。

（1）public Socket（String host，int port）：创建一个流连接，host指定主机，port指定端口号。

（2）public Socket（InetAddress address，int port）：创建一个流连接，host指定Internet地址，port指定端口号。

（3）public Socket（String host，int port，boolean stream）：创建一个Socket类，host指定主机，port指定端口号，stream为true时Socket连接为流连接，否则为数据报连接。

（4）public Socket（InetAddress host，int port，boolean stream）：创建一个Socket类，host指定Internet地址，port指定端口号，stream为true时Socket连接为流连接，否则为数据报连接。

### 2. Socket类的常用方法

Socket类的常用方法如表12－3－1所示。

表 12-3-1　Socket 类的常用方法

| 方法 | 功能描述 |
|---|---|
| int getPort() | 获取端口号 |
| int getLocalPort() | 获取本地端口号 |
| boolean isClosed() | 判断 Socket 类是否关闭，关闭则返回 true，否则返回 false |
| InetAddress getInetAddress() | 返回被连接的服务器地址 |
| InetAddress getLocalAddress() | 获取本地地址 |
| InputStream get InputStream() | 获取 Socket 输入流 |
| OutputStream get OutputStream() | 获取 Socket 输出流 |
| void shutdownInput() | 关闭输入流 |
| boolean isInputShutdown() | 判断输入流是否关闭，若关闭则返回 true，否则返回 false |
| boolean isConnected() | 判断 Socket 类是否被连接，若被连接则返回 true，否则返回 false |

## 二、ServerSocket 类

### 1. ServerSocket 类的构造方法

ServerSocket 类在服务器端使用，其常用构造方法如下。

（1）ServerSocket（int port）：创建一个 ServerSocket 类，port 指定端口号。

（2）ServerSocket（int port，int queueLength）：创建一个 ServerSocket 类，port 指定端口号，queueLength 指定并发等待时队列的最大客户数。

### 2. ServerSocket 类的常用方法

ServerSocket 类的常用方法如表 12-3-2 所示。

表 12-3-2　ServerSocket 类的常用方法

| 方法 | 功能描述 |
|---|---|
| Socket accept() | 等待客户端请求，若连接，则返回一个 Socket 类对象 |
| void close() | 关闭服务器端的 Socket 连接 |
| boolean isClosed() | 判断服务器端的 Socket 连接是否关闭，若关闭则返回 true，否则返回 false |
| InetAddress get InetAddress() | 返回与服务器端的 Socket 连接的 IP 地址 |
| int getLocalPort() | 获取服务器端的 Socket 的端口号 |
| void bind(SocketAddress endpoint) | 绑定 Socket 的 IP 地址，endpoint 指定 IP 地址和端口号 |
| boolean isBound() | 判断服务器端的 Socket 是否与某个 Socket 的 IP 地址绑定，若已绑定则返回 true，否则返回 false |

## 三、Socket 通信工作过程

服务器端监听某个端口是否有连接请求，当客户端向服务器端发出连接请求时，若服务器端监听到该请求，则向客户端回送一个接收连接的消息，连接建立后，双方均可通过 send()、write()等方法与对方进行通信，如图 12-3-1 所示。具体实现步骤如下。

（1）创建 Socket。

（2）建立连接到 Socket 的 IO 流。

（3）对 Socket 进行读写操作。

（4）关闭 Socket 连接。

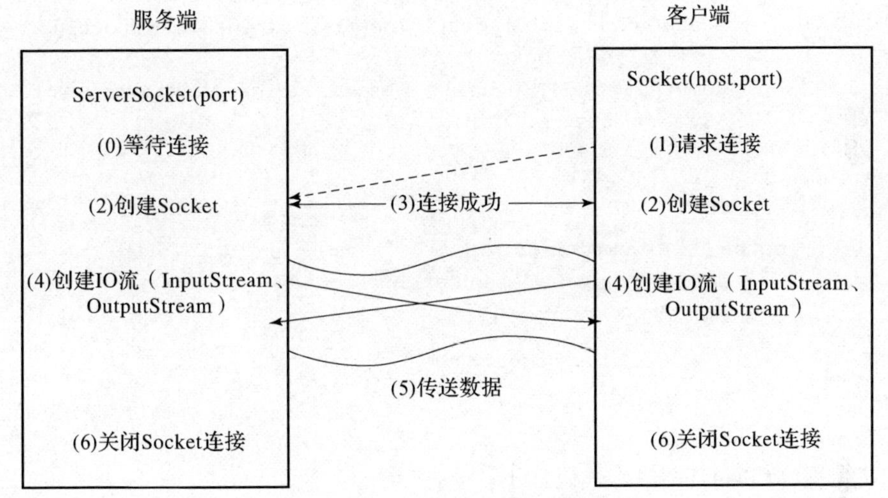


图 12-3-1　Socket 通信工作过程

【例 12-3-1】服务器端与客户端使用 Socket 建立连接，操作界面如图 12-3-2 所示。首先，必须先单击窗体中“启动服务器”按钮，启动服务器端的 Socket；其次，单击“请求连接服务器端，接收服务器端数据”按钮，新建一个客户端的 Socket 连接服务器端的 Socket，接收服务器端发送的数据“hello，i am server，can you receive?”并输出。由于客户端程序与服务器端程序必须同时启动，所以使用线程实现。

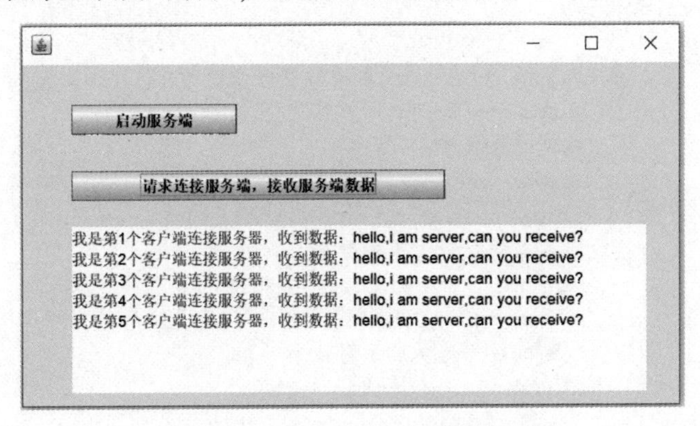


图 12-3-2　Socket 应用操作界面

服务器端线程（ServerThread）代码如下：

```
1   public class ServerThread extends Thread{
2       @ Override
3       public void run() {
4           try {
5               //新建服务器端的 Socket ,端口号 3456
6               ServerSocket ss = new ServerSocket(3456);
7               //永真循环实现实时监测
8               while (true ) {
9                   //服务器端的 Socket 等待连接请求
10                  Socket s1 = ss.accept();
11                  //服务器端创建输出流,向客户端的 Socket 发送数据
12                  OutputStream out1 = s1.getOutputStream();
13                  DataOutputStream dout1 = new DataOutputStream(out1);
14                  //向客户端的 Socket 发送数据
15                  dout1.writeUTF("hello,i am server,can you receive?");
16                  System. out. println("客户端已连接");
17                  //关闭连接
18                  s1.close();
19              }
20          } catch (IOException e) {
21              //TODO Auto - generated catch block
22              e.printStackTrace();
23          }
24      }
25  }
```

客户端线程（ClientThread）代码如下：

```
1   public class ClientThread extends Thread{
2       //声明窗体,并提供 set()方法
3       SocketFrame sf;
4       public void setSf(SocketFrame sf) {
5           this .sf = sf;
6       }
7       @ Override
8       public void run() {
9           try {
10              //新建客户端的 Socket,申请与服务器端的 3456 端口连接
11              Socket s2 = new Socket("localhost",3456);
12              //客户端创建输入流
13              InputStream in2 = s2.getInputStream();
14              DataInputStream din2 = new DataInputStream(in2);
15              //接收服务器端发送过来的数据
16              String getmessage = din2.readUTF();
17              System. out. println("我是客户端,我收到的数据:" + getmessage);
18              //将接收到的数据传回展示窗体
19              sf.setGetmessage(getmessage);
```

```
20              s2.close();
21          } catch (IOException e) {
22              //TODO Auto - generated catch block
23              e.printStackTrace();
24          }
25      }
26  }
```

窗体代码如下：

```
1  public class SocketFrame extends JFrame {
2      private static final long serialVersionUID = 1L;
3      private JPanel contentPane;
4      private JTextArea textArea;
5      private String getmessage;
6      //客户端计数器
7      int i = 0;
8      //展示结果字符串
9      String result = "";
10     //接收客户端线程传回数据,并在多行文本框显示结果
11     public void setGetmessage(String getmessage) {
12         i ++;
13         this.getmessage = "我是第" + i + "个客户端连接服务器,收到数据:" + getmessage + "\n";
14         result = result + this.getmessage;
15         textArea.setText(result);
16     }
17     public static void main(String[] args) {
18         EventQueue.invokeLater(new Runnable() {
19             public void run() {
20                 try {
21                     SocketFrame frame = new SocketFrame();
22                     frame.setVisible(true);
23                 } catch (Exception e) {
24                     e.printStackTrace();
25                 }
26             }
27         });
28     }
29     public SocketFrame() {
30         setDefaultCloseOperation(JFrame.EXIT_ON_CLOSE);
31         setBounds(100, 100, 535, 300);
32         contentPane = new JPanel();
33         contentPane.setBorder(new EmptyBorder(5, 5, 5, 5));
34         setContentPane(contentPane);
35         contentPane.setLayout(null);
36 
37         JButton btnNewButton = new JButton("启动服务器端");
38         btnNewButton.addActionListener(new ActionListener() {
39             public void actionPerformed(ActionEvent e) {
```

```
40                    ServerThread serverThread = new ServerThread();
41                    serverThread.start();
42                }
43            });
44            btnNewButton.setBounds(39, 31, 131, 23);
45            contentPane.add(btnNewButton);
46
47            JButton btnNewButton_1 = new JButton("请求连接服务器端,接收服务器端数据");
48            btnNewButton_1.addActionListener(new ActionListener() {
49                public void actionPerformed(ActionEvent e) {
50                    //创建客户端线程
51                    ClientThread clientThread = new ClientThread();
52                    //将本窗体发送至客户端线程,等待客户端线程回传数据
53                    clientThread.setSf(SocketFrame.this);
54                    //启动客户端程序
55                    clientThread.start();
56                }
57            });
58            btnNewButton_1.setBounds(39, 82, 296, 23);
59            contentPane.add(btnNewButton_1);
60            textArea = new JTextArea();
61            textArea.setBounds(39, 125, 455, 128);
62            contentPane.add(textArea);
63        }
64    }
```

任务工单

| 任务名称 | 图书信息管理系统——留言板 | | | | |
|---|---|---|---|---|---|
| 班级 | | 学号 | | 姓名 | |
| 任务要求 | （1）启动服务器端，接收留言并显示。<br>（2）多客户端向服务器端发送留言。 | | | | |
| 任务实现 | | | | | |
| 异常及待解决问题记录 | | | | | |

# 12.4 HTTP 通信及 JSON

## 学习目标

（1）了解 HTTP 通信原理。

（2）掌握字符串类的初始化与使用方法。

（3）掌握 String 类、StringBuilder 类、StringBuffer 类的相关操作方法。

## 相关知识

### 12.4.1 HTTP 通信

HTTP 是一种基于请求-响应模型的简单通信协议，它运行在 TCP 之上。HTTP 定义了客户端和服务器端之间交换信息的方式，客户端可以通过发送 HTTP 请求获取服务器端的资源，而服务器端则以 HTTP 应答的形式返回这些资源。HTTP 是无状态的协议，这意味着每个请求和应答都是独立的，客户端和服务器端之间不需要维持持久的连接。每个连接在完成请求和应答后都会关闭。

#### 一、HTTP 请求方式

HTTP 通信最常用的两种方式为 POST 方式和 GET 方式。POST 方式通过 HTTP 消息实体发送数据给服务器端，安全性高，数据传输大小没有限制，常用于注册、登录等安全性较高且向数据库中写入数据的操作。GET 方式通过 URL 的查询字符串向服务器端传递参数，以明文显示在浏览器地址栏，保密性低，最多传输 2 048 个字符，但 GET 方式的请求效率高，多用于查询（读取资源）。常用的 HTTP 请求方式如表 12-4-1 所示。

表 12-4-1 常用的 HTTP 请求方式

| HTTP 请求方式 | 功能描述 |
| --- | --- |
| GET | 用于从指定资源请求数据 |
| POST | 用于向指定资源提交数据进行处理请求 |
| HEAD | 类似 GET 方式，但不会返回具体的响应内容，仅返回状态行和标题信息 |
| PUT | 用于将数据发送到服务器端以创建或更新资源 |
| DELETE | 用于删除指定的资源 |
| OPTIONS | 用于询问服务器端支持哪些 HTTP 请求方式 |
| TRACE | 用于追踪服务器端收到的请求，主要用于测试或诊断 |
| CONNECT | 用于建立到给定 URI 标识的服务器端的通道 |
| PATCH | 用于部分更新指定资源的数据 |

## 二、HTTP 响应

### 1. HTTP 响应格式

HTTP 相应格式如图 12－4－1 所示。

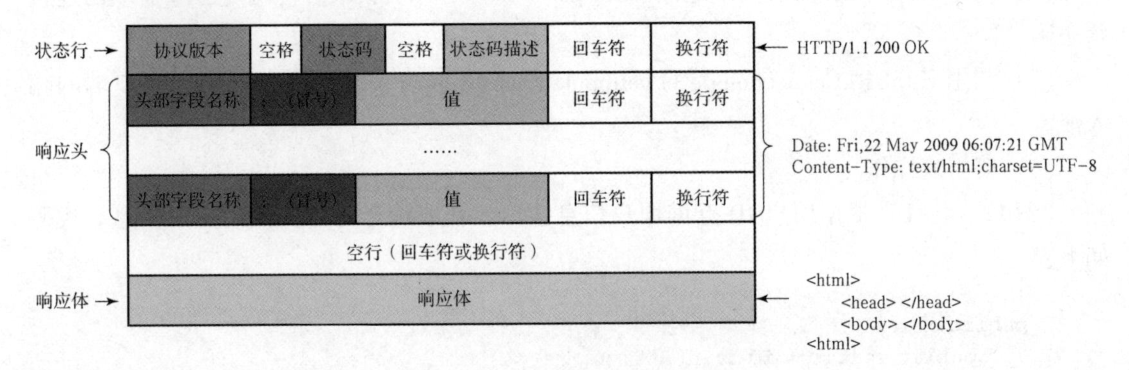


图 12－4－1　HTTP 响应格式

（1）状态行：协议版本（空格）状态码（空格）状态描述。

（2）响应头。

①语法格式：key：value。

②Content－Type：描述响应体中数据类型。

（3）空行：表示响应头结束。

（4）响应体：若请求成功，则发回数据；若请求失败，则发回错误信息。

### 2. 状态码

HTTP 状态码由 3 个十进制数字组成，第 1 个十进制数字表示状态码的类型，后 2 个十进制数字表示对状态码的细分。常见的状态码如表 12－4－2 所示。

表 12－4－2　常见的状态码

| 状态码 | 状态英文名称 | 描述 |
| --- | --- | --- |
| 200 | OK | 请求成功，一般用于 GET 和 POST 请求 |
| 400 | Bad Request | 语义有误，当前请求无法被服务器端理解，或者请求参数有误 |
| 404 | Not Found | 请求失败，服务器端无法根据客户端请求找到资源 |
| 500 | Internal server Error | 服务器端遇到不知道如何处理的情况，一般为程序出错 |

## 三、HttpURLConnection 类

HttpURLConnection 类继承 URLConnection 类，位于 java. net 包中，它对外提供访问 HTTP 的基本功能，可用于向指定网站发送 GET 请求、POST 请求。HttpURLConnection 类的使用步骤如下。

（1）实例化 URL 类对象。

（2）调用 URL 类对象的 openConnection( ) 方法创建 HttpURLConnection 类对象，通过 HttpURLConnection 类对象建立客户端与服务器端的连接。此处需要将 openConnection( ) 方法返回的 URLConnection 类对象强制转换为 HttpURLConnection 类型。

（3）调用 HttpURLConnection 类的 getResponseCode( ) 方法获取客户端与服务器端的连接状态码。

（4）调用 HttpURLConnection 类的 getInputStream( ) 方法获取从服务器端到客户端的输入流。

（5）解析流操作。

**【例 12-4-1】**根据用户 ID 查询用户信息接口，获取编号为 001 的用户的信息。代码如下：

```
1  public class P12_5 {
2      public static void main(String[] args) {
3          try {
4              //定义访问的网络资源
5              URL url = new URL("http://101.34.44.3:8080/javahttp/rest/
                                        selectUserById/001");
6              //创建于资源的连接
7              HttpURLConnection conn = (HttpURLConnection)url.openConnection();
8              //设置程序允许使用 InputStream 从服务器端读入数据
9              conn.setDoInput(true);
10             //设置请求方式为 GET
11             conn.setRequestMethod("GET");
12             //设置通用请求参数
13             conn.setRequestProperty("Accept",
                               "application/json;charset=UTF-8");
14             conn.setRequestProperty("Content-Type",
                               "application/json;charset=UTF-8");
15             //获取 HttpURLConnection 输入流
16             InputStream is = conn.getInputStream();
17             InputStreamReader isr = new InputStreamReader(is);
18             BufferedReader br = new BufferedReader(isr);
19             String line = "";
20             //循环读入数据
21             while ((line = br.readLine()) != null) {
22                 System.out.println(line);
23             }
24             conn.disconnect();
25         } catch (Exception e1) {
26             e1.printStackTrace();
27         }
28     }
29 }
```

运行结果如图 12-4-2 所示。

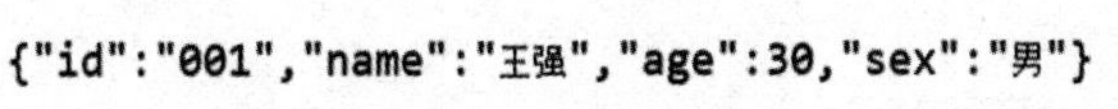

图 12-4-2　运行结果

## 12.4.2　JSON 文本格式

JavaScript 对象表示法（JavaScript Object Notation，JSON）是一种轻量级的数据交换格式，它以易于阅读和编写的文本格式表示结构化数据，通常用于 Web 应用程序之间的数据传输。

### 一、JSON 语法规则

（1）数据以名称/值的形式表示，形如“" 名称"：值”。

（2）数据由逗号分隔。

（3）大括号保存对象，中括号保存数组。

例如：

```
//JSON 对象
{"name":"Tom","age":20,"sex":true}
//JSON 数组
[
      {"name":"Tom","age":20,"sex":true},
      {"name":"Jerry","age":18,"sex":false}
]
```

### 二、Java 中 JSON 的使用

Java 常用的第三方 JSON 处理库有阿里巴巴开发的 FastJson、谷歌开发的 Gson、美国 FasterXML 公司的 Jackson，本书使用谷歌的 Gson 库展开 JSON 的学习。学习前，先在工程中导入 Gson 相应的 jar 包“gson-2.3.1.jar”。Gson 的常用方法如下。

（1）将 JSON 字符串转换为对象，语法格式如下：

```
fromJson(String json,Class<T> classOfT)
```

（2）将对象转换为 JSON 字符串，语法格式如下：

```
toJson(Object obj)
```

**【例 12-4-2】**

```
1  import com.google.gson.Gson;
2  import com.google.gson.reflect.TypeToken;
3  public class P12_7 {
4      public static void main(String[] args) {
5          //JSON 对象,注意双引号的转义
6          String json1 = "{\"name\":\"Tom\",\"age\":20,\"sex\":true}";
7          //JSON 数组
8          String json2 = "[{\"name\":\"Tom\",\"age\":20,\"sex\":true},
                          {\"name\":\"Jerry\",\"age\":18,\"sex\":false}]";
```

```
9        Gson gson = new Gson();
10         //JSON 字符串转换为 User 对象
11         User user = gson.fromJson(json1, User.class);
12         System.out.println("json1 转为 User 对象:" + user);
13         //JSON 字符串转换为 User 集合
14         List<User> users = gson.fromJson(json2,
                      new TypeToken<List<User>>(){}.getType());
15         System.out.println("json2 转为 User 集合:" + users);
16         User u1 = new User();//User 对象
17         u1.setName("Joe");
18         u1.setAge(10);
19         u1.setSex(true);
20         User u2 = new User();
21         u2.setName("Mary");
22         u2.setAge(12);
23         u2.setSex(false);
24         List<User> userlist = new ArrayList<User>();//User 集合
25         userlist.add(u1);
26         userlist.add(u2);
27         String json3 = gson.toJson(u1);
28         String json4 = gson.toJson(userlist);
29         System.out.println("User 对象转换为 JSON 字符串:" + json3);
30         System.out.println("User 集合转换为 JSON 字符串:" + json4);
31     }
32 }
```

运行结果如图 12-4-3 所示。

```
json1转换为User对象：User [name=Tom, age=20, sex=true]
json2转换为User集合：[User [name=Tom, age=20, sex=true], User [name=Jerry, age=18, sex=false]]
User对象转换为JOSN字符串：{"name":"Joe","age":10,"sex":true}
User集合转换为JOSN字符串：[{"name":"Joe","age":10,"sex":true},{"name":"Mary","age":12,"sex":false}]
```

图 12-4-3　运行结果

**案例**

使用“客户信息管理系统”网站接口，实现客户信息查询功能。

**1. 接口**

（1）根据客户姓名模糊查询客户信息接口（GET 请求，参数为一个客户姓名，返回值为一个 Complex 类对象），网址为 http://101.34.44.3:8086/v1/customer/getName? name = {1}。

（2）根据电话号码精确查询客户信息接口（GET 请求，参数为一个客户电话号码，返回值为一个 Complex 类对象），网址为 http://101.34.44.3:8086/v1/customer/getTel? tel = {13600987987}。

**2. 封装类设计**

（1）客户类代码如下。

```
public class Customer {
    private int cid;
    private String name,sex,tel,password;
    private int rewardpoints;
    //省略访问方法、设置方法、构造方法、toString()方法
}
```

(2) 接口返回值类代码如下：

```
public class Complex {
    private int code;
    private String msg;
    private Object data; //返回值可能是一个集合,也可能是一个类对象
    //省略访问方法、设置方法、构造方法、toString()方法
    }
```

### 3. 窗体设计

窗体设计如图 12 -4 -4 所示。

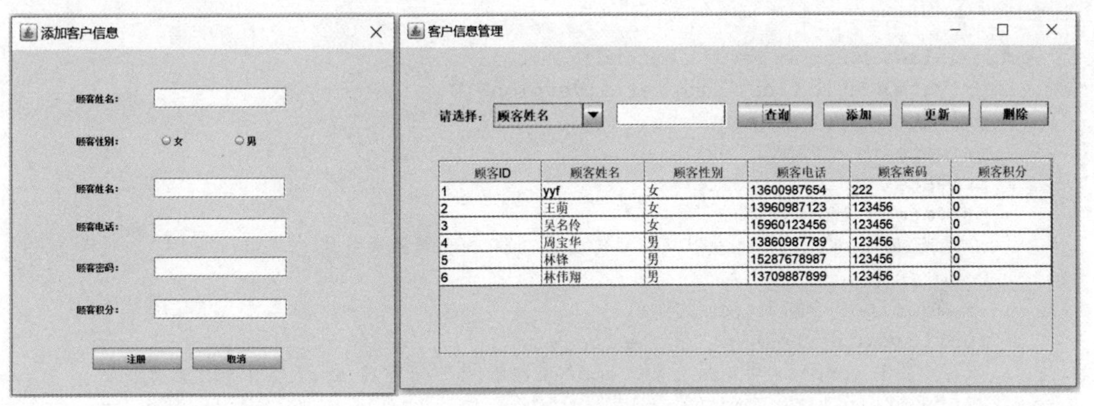


图 12 -4 -4　窗体设计

### 4. 工具类“CustomerFunction. java”

代码如下：

```
public class CustomerFunction {
    public static void selectCustomer(String key,String value,
                                      CustomerDetail customerDetail) {
        if (key.equals("顾客姓名")) {
            SelectCustLikeNameThread selectCustLikeNameThread = new
                                              SelectCustLikeNameThread();
            selectCustLikeNameThread.setValue(value);
            selectCustLikeNameThread.setCustomerDetail(customerDetail);
            selectCustLikeNameThread.start();
        }
    }
    public static Customer getSelectedCustomer(JTable table,int n) {
```

```
            String cid = table.getValueAt(n, 0).toString();
            String name = table.getValueAt(n, 1).toString();
            String sex = table.getValueAt(n, 2).toString();
            String tel = table.getValueAt(n, 3).toString();
            String password = table.getValueAt(n, 4).toString();
            String rewardpoints = table.getValueAt(n, 5).toString();
            Customer c = new Customer();
            c.setCid(Integer.parseInt(cid));
            c.setName(name);
            c.setSex(sex);
            c.setTel(tel);
            c.setPassword(password);
            c.setRewardpoints(Integer.parseInt(rewardpoints));
            return c;
        }
    }
```

5. 功能实现

（1）窗体交互功能，代码如下：

```
public class CustomerDetail extends JFrame {
    private static final long serialVersionUID = 1L;
    private JPanel contentPane;
    private JTextField textField;
    private JComboBox comboBox;
    private JTable table;
    //定义 Complex 类对象,获取从线程传回的 Complex 类结果对象
    Complex result;
    //本方法,客户的增、删、改操作调用
    public void setResult(Complex result) {
        //获取 Complex 类结果对象的 code 属性值,为 1 表示增、删、改操作执行成功
        if (result.getCode() ==1)
            JOptionPane.showMessageDialog(null , result.getMsg());
        else
            JOptionPane.showMessageDialog(null , result.getMsg());
        String key = comboBox.getSelectedItem().toString();
        String value = textField.getText();
        //调用工具类 CustomerFunction 的 selectCustomer()方法重新加载展示数据
        CustomerFunction.selectCustomer(key, value, CustomerDetail. this );
    }
    //本方法,查询客户信息相关操作调用
    public void setResult1(List < Customer > custlist) {
        //遍历从线程查询返回的客户信息集合 List < Customer >
        Iterator < Customer > ite = custlist.iterator();
        //将 List < Customer > 转换为 Vector 集合,为创建表格模型提供数据
        Vector data = new Vector();
        while (ite.hasNext()){
            Customer c = ite.next();
            Vector row = new Vector();
```

```
            row.add(c.getCid());
            row.add(c.getName());
            row.add(c.getSex());
            row.add(c.getTel());
            row.add(c.getPassword());
            row.add(c.getRewardpoints());
            data.add(row);
        }
        Vector columnNames = new Vector();
        columnNames.add("顾客ID");
        columnNames.add("顾客姓名");
        columnNames.add("顾客性别");
        columnNames.add("顾客电话");
        columnNames.add("顾客密码");
        columnNames.add("顾客积分");
        TableMode mode = new TableMode(data, columnNames);
        table.setModel(mode);
    }
    ……(略)
    public CustomerDetail() {
        ……(略)
        JButton btnNewButton  = new JButton("查询");
        btnNewButton.addActionListener(new ActionListener() {
            public void actionPerformed(ActionEvent e) {
                String key = comboBox.getSelectedItem().toString();
                String value = textField.getText();
                //查询关键字为"顾客姓名",则调用SelectCustLikeNameThread线程
                if (key.equals("顾客姓名")) {
                    SelectCustLikeNameThread selectCustLikeNameThread = new
                                                SelectCustLikeNameThread();
                  selectCustLikeNameThread.setValue(value);
                  //将当前窗体传到线程,以便将查询结果返回本窗体展示
                  selectCustLikeNameThread.setCustomerDetail(
                                                CustomerDetail.this);
                  selectCustLikeNameThread.start();
                }
               //查询关键字为"顾客姓名",则调用SelectCustByIdThread线程
            if (key.equals("顾客电话")) {
                  SelectCustByIdThread selectCustByIdThread = new
                                                SelectCustByIdThread();
                  selectCustByIdThread.setTel(value);
                  //将当前窗体传到线程,以便将查询结果返回本窗体展示
                  selectCustByIdThread.setCustomerDetail(
                                                CustomerDetail.this);
                  selectCustByIdThread.start();
                }
            }
        });
        ……(略)
    }
}
```

（2）模糊查询线程，代码如下：

```
public class SelectCustLikeNameThread extends Thread{
    HttpURLConnection conn;
    String json = "";
    //定义 CustomerDetail 窗体,接收查询结果并展示
    CustomerDetail customerDetail;
    public void setCustomerDetail(CustomerDetail customerDetail) {
        this.customerDetail = customerDetail;
    }
    String value; //定义 value,获取要查询的姓名
    public void setValue(String value) {
    this.value = value;
  }
  public void run() {
      try {
          //访问根据姓名模糊查询顾客信息接口
          URL url = new URL("http://101.34.44.3:8086/v1/customer/getName?
                      name = ".concat(URLEncoder.encode(value, "UTF-8")));
          //创建网络连接
          conn = (HttpURLConnection)url.openConnection();
          conn.setDoInput(true);
          conn.setDoOutput(true);
          conn.setRequestMethod("GET");
          conn.setRequestProperty("Accept", "application/json;charset =
                                                              UTF-8");
          conn.setRequestProperty("Content-Type","application/json;
                                                 charset = UTF-8");
          conn.setConnectTimeout(5000);
          //生成从接口获取数据的输入流
          InputStream is = conn.getInputStream();
          InputStreamReader isr = new InputStreamReader(is);
          BufferedReader br = new BufferedReader(isr);
          String line = "";
          //循环读取接口传回的 JSON 字符串
          while ((line = br.readLine())! = null){
              json = json.concat(line).concat("\n");
          }
          System.out.println("json:" + json);
    } catch (Exception e1) {
          e1.printStackTrace();
    }finally {
        if (conn! = null){
            conn.disconnect();
        }
    }
    Gson gson = new Gson();
    //将传回的 JSON 字符串转换为 Complex 类对象
    Complex complex = gson.fromJson(json, Complex.class);
    //获取 Complex 类对象中客户信息的 JSON 字符串,此时收到的是一个客户集合
    String datajson = complex.getData().toString();
    //将包含客户信息的 JSON 字符串转换为 List<Customer>集合
    List<Customer> custlist = gson.fromJson(datajson, new
                        TypeToken<List<Customer>>(){}.getType());
    //将结果集传回窗体展示
    customerDetail.setResult1(custlist);
  }
}
```

## 任务工单

| 任务名称 | 图书信息管理系统——更新图书功能 | | | | |
|---|---|---|---|---|---|
| 班级 | | 学号 | | 姓名 | |

**任务要求**

(1) 接口。

①根据图书编号查询图书信息接口：http://101.34.44.3:8086/v1/book/getId?id={1}。

②更新图书接口：http://101.34.44.3:8086/v1/book/update。

(2) 接口返回值封装类代码如下：

```
public class Complex {
        private int code;
        private String msg;
        //返回值可能是一个集合,也可能是一个类对象
        private Object data;
        //添加访问方法、设置方法
        //添加构造方法
        //添加 toString()方法
}
```

(3) 输入图书编号，查询图书信息，并显示在文本框中。

(4) 修改图书信息，单击“更新图书信息”按钮，完成图书更新。

续表

<table>
<tr><td>任务名称</td><td colspan="5">图书信息管理系统——更新图书功能</td></tr>
<tr><td>班级</td><td></td><td>学号</td><td></td><td>姓名</td><td></td></tr>
<tr><td>任务实现</td><td colspan="5">(提交窗体交互代码、线程代码)</td></tr>
<tr><td>异常及待解决<br>问题记录</td><td colspan="5"></td></tr>
</table>

## 练习题

### 1. 选择题

(1) InetAddress 类中能获取 IP 地址及主机名的方法是（　　）。

A. getHostName()　　B. getLocalHost()

C. getHostAddress()　　D. getAddress()

(2) 在 UDP 通信中，用于发送数据包的类是（　　）。

A. DatagramPacket　　B. ServerSocket

C. Socket　　D. DatagramSocket

(3) 下列 ServerSocket 类的方法中，用于接收来自客户端请求的方法是（　　）。

A. accept()　　B. getOutputStream()

C. receive()　　D. get()

(4) 在 Java 网络编程中，使用客户端套接字 Socket 创建对象时，需要指定（　　）。

A. 服务器名和端口　　B. 服务器端口和文件

C. 服务器名和文件　　D. 服务器地址和文件

(5) ServerSocket 类的监听方法 accept()的返回值类型是（　　）。

A. Socket　　B. Void

C. Object　　D. DatagramPacket

(6) 一个服务器进程执行以下代码：

```
ServerSocket serverSocket = new ServerSocket(80);
Socket socket = serverSocket.accept();
int port = socket.getPort();
```

以下说法不正确的是（　　）。

A. 服务器进程占用 80 端口

B. socket. getPort()方法返回服务器进程所占用的本地端口，此处返回值是 80

C. serverSocket. accept()方法成功返回，表明服务器进程接收到一个客户连接请求

D. socket. getPort()方法返回客户端 Socket 所占用的本地端口

(7) HTTP 规定，在默认情况下，HTTP 服务器占用的 TCP 端口号是（　　）。

A. 21　　B. 80

C. 23　　D. 任意一个未被占用的端口号

### 2. 填空题

(1) 使用 UDP 开发 Web 应用程序时，需要使用两个类，分别是________和________。

(2) 网络通信协议有很多种，目前应用最广泛的是________、________、________和其他一些协议的协议组。

(3) 统一资源定位符 URL 是指向互联网资源的指针，它由 4 个部分组成：协议名、主机名、________和资源路径标识。

(4) 在 Socket 编程中，IP 地址用来标识一台计算机，但是一台计算机中可能有多种应

用程序，使用 ______ 区分这些应用程序。

（5）UDP 称为____________协议，即在数据传输时，数据的发送端和接收端不建立__________。

（6）JDK 提供了两个用于实现 TCP 程序的类，一个是________类，用于表示服务器端；另一个是________ 类，用于表示客户端。

**3. 编程题**

使用基于 TCP 的 Java Socket 编程，实现如下功能。

（1）从客户端输入几个字符，发送到服务器端。

（2）由服务器端将接收到的字符输出。

（3）服务器端向客户端发出“您的信息已收到”作为响应。

（4）客户端接收服务器端的响应信息。